REVOLUTIONIZING THE SCIENCES

REVOLUTIONIZING THE SCIENCES

European Knowledge and Its Ambitions, 1500–1700

Peter Dear

Princeton University Press
Princeton, New Jersey

Published in the United States and Canada by Princeton University Press,
41 William Street, Princeton, New Jersey 08540

Published in North America under license from Palgrave, Houndmills, Basingstoke,
Hants RG21 6XS, United Kingdom

First published 2001 by PALGRAVE

Library of Congress Catalog Card Number 00-109720

ISBN 0-691-08859-4 (cloth)
ISBN 0-691-08860-8 (paperback)

This book has been composed in Palatino

Printed on acid-free paper. ∞

www.pupress.princeton.edu

Printed in the United States of America

10   9   8   7   6   5   4   3

Contents

Preface

This is a book intended for use by undergraduates in connection with college and university courses, as well as by others interested in gaining an overview of what is usually called "the Scientific Revolution." As such, its chief purpose is to provide a framework suitable for facilitating more intensive study of the multifarious issues that arise from the narrative provided here. The bibliographical discussion at the end points the way to much important scholarly literature regarding many of the more prominent such issues.

At the same time, a book of this kind cannot cover everything (or, indeed, anything) adequately. It is my hope that it will, at least, suggest to readers topics deserving of closer investigation and avenues by which to investigate them. Overall, the book cleaves fairly closely to a view of the period that should be broadly familiar to university teachers of relevant courses; if it did not, it would be of little use to them or their students. Thus there is a stronger focus on mathematical and physical sciences than on life sciences or medicine. The latter are, indeed, discussed throughout the book, but there is a strong case to be made (and, by others, denied) that the most significant intellectual developments of the sciences in the period reviewed occurred in areas of methodology, matter theory, and mathematical sciences; thus, for example, my discussion of natural history, while an important component of the overall argument, is limited in its technical content (as, to be sure, are most considerations of mathematical sciences).

Similarly, I have been obliged to deal with the relevant social history of the period primarily when it intersects directly with discussion of institutional and conceptual matters concerning the study of nature by the learned élite: more extensive consideration of, for example, gender issues in the formation of modern science in this period, or with issues of class (the latter obviously crucial but, as yet, under-researched) could not be carried out within the limits of an introduction of this sort, but I have tried to provide pointers in the text to their potential significance. Once again, these are all issues that can be pursued further by following leads included in the bibliographical essay.

I would like to thank the anonymous reviewers of this book, and particularly Paula Findlen, for extremely useful comments on the manuscript, which have improved it significantly. Alas, I must take the blame for the faults that remain.

The book has been written with the intention that it be used in concert with associated primary-source material in English translation. The editions cited in the notes to individual chapters would make valuable study materials to accompany this book's overall narrative.

PETER DEAR

Introduction
Philosophy and Operationalism

I Knowledge and its history

What is knowledge? A bird, we say, knows how to fly. But we would not like to claim that it therefore knows aeronautics: there have never been avian Wright brothers.

There is much invested in the word "knowledge," and as with any word that bears many connotations, this one has a long and complex history. An understanding of the meanings that it carries for us today will therefore require a journey into the regions of the past where those meanings were first created in a recognizably modern form. One of the most important is the Europe of the sixteenth and seventeenth centuries, a time and place that, in the history of science, is usually known as the Scientific Revolution.

The global practice that we call science is still, in the twenty-first century, coordinated with primary reference to centres of training and research that look to the European tradition. This tradition was first adopted elsewhere on a large scale in the United States, often with the help of European training and European émigrés, and only in the twentieth century did it become naturalized elsewhere. Nobel prizes in the sciences even now go predominantly to scientists in Europe and North America, including scientists from elsewhere in the world who received their training and conducted their research in those places. An historical understanding of that characteristically modern enterprise must therefore look first to its development in a European setting.

The idea that something particularly important to the emergence of European science occurred in the sixteenth and seventeenth centuries is one that Europeans themselves first claimed in the eighteenth century. The period from the work of Copernicus in the early sixteenth century, which put the earth in motion around the sun, up to the establishment of the Newtonian world-system at the start of the eighteenth – which included universal gravitation as part of an indefinitely large universe – came to be regarded

1

as a marvellous "revolution" in knowledge unparalleled in history.[1] Naturally, this perspective included an appropriate evaluation of what had gone before. The European learning of the Middle Ages, on this view, had been backward and empty. Philosophers had been slaves to the ancient writings of Aristotle; they had been more concerned with words and arguments than with things and applications. It is a view that still lives on in popular myth, despite the radical historical re-evaluations of the Middle Ages, accomplished during the past century, that have given the lie to such a dismissive caricature of medieval intellectual life. Nonetheless, some aspects of the eighteenth century's celebratory account of its recent forebears deserve continued attention. For all that it was exaggerated and self-congratulatory, the idea that there was a fundamental difference between medieval learning and the new learning brought about by the recent "revolution" contains an important insight. Medieval learning, on this account, had stressed the ability to speak about matters of truth; whereas now, instead, there was a stress on knowledge of what was in the world and what it could do.

This book will, in effect, examine how much justice that view contains. The story will be more complicated than the easy triumphalist accounts of the eighteenth century, however. We are nowadays less confident than the spokesmen of the Enlightenment that there had been an unambiguous triumph of rationality over obfuscation, or that our own modern science is a neutral and inevitable product of progress. That science is a part of the culture that nurtures it has been shown time and again by so-called "contextualist" historical and sociological studies of specific cases; science, they have shown, is made by history. The central goal of the history of science is to understand why particular people in the past believed the things they did about the world and pursued inquiries in the ways they did. The historian has no stake in adjudicating the truth of past convictions. No historical understanding of Copernicus's belief in the motion of the earth around the sun comes from the proposition that his belief *was true*. Copernicus believed what he did for various *reasons*, which it is the job of the historian to find out; truth or falsity are determined by arguments, and it is the arguments that can be studied historically.

In explaining historical change, many factors may be invoked, often different ones in different cases. A difficulty in historical work arises from its complexity and the frequent singularity of the events or situations being addressed. It is as if a geologist were to be called upon to explain why a particular mountain happened to be exactly as high as it was, no more and no less. The elevation of such mountains might be explicable in terms of general geological processes, but the exact details of the appearance of any particular one would be too much dependent on the unknown, accidental contingencies of its history. Historians, similarly, cannot provide deductive causal accounts of why a particular event, such as the English Civil War, took place in the way that it did. They can attempt, however, to make generalizations about what conditions rendered such an event more or less likely. Another way of seeing this is to move away from talking about the likelihood of outcomes, to speak instead of *understanding*. The historian wants to understand aspects of the past in the same sort of way as we understand what was involved in our neighbour's winning the lottery, even though we could not have predicted it.

In the Scientific Revolution, similar issues were at stake for investigators of nature. Their medieval predecessors, destined to be pilloried in the eighteenth century, had aimed above all at *understanding* the natural world; the new philosophers typically aimed, by contrast, at successful prediction and control. It was not a matter of doing the same thing better – it was a matter of doing something *different*. The literate culture of the High Middle Ages (roughly, the twelfth century to the fourteenth century) had grown up around the medieval universities, in which it was generally known as "scholasticism." These new institutions were to a greater or lesser degree associated with the church and with its cultural agenda. As a result, at universities such as those of Paris or Oxford theology was the first among their higher faculties (those granting the doctorate); it was routinely known as "the queen of the sciences." Scholarly prestige tended as a result to accrue to abstract philosophizing intended to serve the establishment of truth; this was the rational counterpart of belief, and spoke to intellectual conviction rather than practical know-how.

The central discipline concerned with knowledge of nature was called "natural philosophy" (*philosophia naturalis* or, often, *scientia naturalis*). Other disciplines also dealt with nature, such as medicine (another of the higher faculties) and the mathematical sciences. These latter, apart from arithmetic and geometry, encompassed studies of those aspects of nature which concerned quantitative properties – areas such as astronomy, music theory, or geometrical optics. Natural philosophy, however, was pre-eminent among all these because it took its central goal to be the philosophical explanation of all aspects of the natural world. It was generally conducted using the relevant writings of Aristotle; because Aristotle had used the Greek word *physis* to refer to the whole of the natural world, living and non-living, the medieval Latin word *physica*, or "physics," was routinely used as a synonym for "natural philosophy."

II How a medieval philosopher thought about the natural world

All revolutions are revolutions against something. One way of doing things is overturned, to be replaced by another, different one. If there really was a scientific "revolution," it must by necessity have overthrown a previous orthodoxy – which is precisely the way the story was told three centuries ago. It is, in fact, unclear to what extent an old, unchallenged orthodoxy had actually existed, or to what extent the ways of thought that replaced it were themselves truly novel and truly unified. But every tale needs a begin-

ning, and the taken-for-granted beliefs of the majority of natural philosophers in the medieval universities provide us with ours. We must therefore
examine the commonplaces of the scholastic-Aristotelian view of natural
knowledge, so that we know a little of what everyone with a university
education knew too.

Aristotelian philosophy was aimed at *explanation*. Aristotle was not
interested in "facts" themselves so much as in what he called the "reasoned
fact." That is, he wanted to know things by knowing *why* things were the
way they were. Mere description of the obvious properties of an object or
process (such as its measureable features) would not, in itself, serve that
explanatory goal; it would merely provide something to be explained. But
this does not mean that the senses, the source of the description, were
devalued. On the contrary, Aristotle had emphasized that all knowledge
ultimately comes by way of the senses. Without the senses, nothing could
be known, not even the truths of mathematics; the latter, like all other items
of knowledge, derived by abstraction from sensory awareness of particulars. The apparently abstract character of medieval Aristotelian philosophy,
the feature most pilloried in the eighteenth century, justified its procedures
by reference to just such a sensory basis. It was not, however, any kind of
experimental ideal that would be recognizable to modern eyes.

To an Aristotelian, sensory knowledge about the world served as the
starting place for the creation of properly *philosophical* knowledge. Consider
the following argument, a standard example in medieval logic:

All men are mortal
Socrates is a man
Therefore Socrates is mortal.

Pieces of sensory information resembled the final line (technically, the
"conclusion" of this "syllogism"): "Socrates is mortal." This is a specific
assertion about Socrates that can be made only on the basis of sensory
experience of that particular person and his actual death. The first line,
however (the "major premise"), that "all men are mortal," is a universal
assertion about *all* men everywhere and at all times. It cannot itself be justified as certain by reference to a delimited set of individual sensory observations. And yet certainty was one of Aristotle's requirements for proper
"scientific" demonstration. During the seventeenth century, critics such as
the Englishman Francis Bacon criticized Aristotelian logical procedures
based on the syllogism for being circular. The universal assertion constituting the major premise could, Bacon said, only be justified on the
basis of countless singulars, of which the conclusion in any given instance
would itself be an example. So the conclusion was being demonstrated on
the basis of a philosophical, universal knowledge-claim that was itself in
part *justified* by the conclusion.[2]

Bacon's criticism should alert us to something unfamiliar in Aristotelian philosophical procedures. Bacon's point was a straightforward one well within the capacities of the enormously logically-sophisticated scholastic philosophers. And yet they did not tend to see it as a meaningful objection. The crucial issue of the move from particular experiences of the world to universally valid (and hence philosophical) generalizations was usually seen as unproblematic. "Experience" for a scholastic Aristotelian did not mean the sensory perception of single events, as might be involved in recording an experimental outcome. Instead, according to Aristotle, "from perception there comes memory, and from memory (when it occurs often in connection with the same thing), experience; for memories that are many in number form a single experience."[3] In effect, Bacon's difficulty is collapsed into a psychological habit; a habit, moreover, that is simply assumed to constitute a legitimate cognitive process. The usual ways in which human beings go about making their knowledge (whether explanatory or inferential) is thus not to be questioned; Aristotle provides a natural history of knowledge rather than a critical epistemology. The Aristotelian position amounts to saying: "If that is what we do, then that is what knowledge is."

Aristotelian experience, in practice, amounted to knowledge that had been gained by someone who had perceived "the same thing" countless times, so as to become thoroughly familiar with it. The rising of the sun every day (making due allowance for cloud-cover) would be an example of such experiential knowledge. That heavy bodies fall downwards was also known to everyone from daily experience, which is why Aristotle could appropriately use it in providing a philosophical explanation of the nature of heavy bodies in his *Physics*.[4] When an Aristotelian philosopher claimed to base his knowledge on sensory experience, he meant that he was *familiar* with the behaviours and properties of the things he discussed. Ideally, his audience would be too. Therein lay the biggest difficulty.

Besides its putative experiential foundations, Aristotelian natural philosophy also claimed to be a *science* (the Latin word used by the scholastics for Aristotle's Greek *epistēmē* was *scientia*). A true science demonstrated its conclusions from premises that were accepted as certain. Demonstrative conclusions would be certain as long as they were deduced correctly from starting points that were themselves certain; mere likelihood was insufficient. This was a very tall order. Aristotle appears to have modelled his conception of an ideal science on the Greek mathematical practice of his day: the kind of geometry exemplified in Euclid's *Elements* (*c.*300 BC) uses as its starting points statements that are taken to be immediately acceptable, being either conventional (definitions) or supposedly self-evident (postulates and axioms). From this foundation, Euclid attempts to derive often unforeseen conclusions regarding geometrical figures by rigorous deduction. Aristotle, in his work *Posterior Analytics*, mandated a similar scheme for all formal bodies of knowledge that aspired to being sciences, regardless of their specific subject-matter. Not surprisingly, Aristotle's ideal found no concrete exemplification outside of Greek mathematics itself. It is difficult to imagine an Aristotelian deductive science of zoology (a field of especial interest to him).

Nonetheless, the lure of demonstrative certainty drew scholastic natural philosophers to believe that they could make knowledge that was analytically solid: terms would be defined in such a way as to permit logically unassailable deductions. Thus, one might define the element "earth" as that substance which has as its natural place the centre of the universe (Aristotle's universe was geocentric). Then, one could easily explain, at least in principle, the centrality of the earthy sphere on which we live (it is where all heavy, meaning earthy, bodies have accumulated), as well as the tendency of heavy bodies to fall downwards (seeking their natural place). It was this kind of explanatory strategy that would look to later critics as a matter of purely verbal trickery.

The Aristotelian reliance on experience that was already universalized (in that sense, already "common knowledge") yielded a natural philosophy that was centrally concerned with explanation *rather than* other goals. The intent was to understand phenomena that were *already known* – there is no pressing sense in which scholastic natural philosophers thought of their enterprise as one of making new *discoveries*. The change in goals represented by the development of such a view is one of the most characteristic features of the large-scale mutations in thought found in the seventeenth century. Discovery itself came most often to be described in geographical terms; in the 1660s, Robert Hooke of the newly-founded Royal Society of London spoke of the microscope as opening up new territories for discovery in the realm of the very small. The expansion of the European perspective brought about by the voyages of discovery to the New World, and the attendant increase in worldwide commerce, made such a metaphor immediately available and accessible. Near the beginning of the seventeenth century, Francis Bacon had made much use of the same image of discovery, and even chose a prophecy from the Book of Daniel to express his programmatic ambitions: "Many shall pass through, and knowledge will be increased."[5] In effect, the world had begun to contain many more things than had been dreamt of in scholastic philosophy.

It is important to recognize, however, that the newly emerging types of natural philosophy that challenged Aristotelianism by the seventeenth century were not simply more efficacious. If they put an increased premium on discovering new things, it was not so clear (to many scholastic-Aristotelians, at least) that they made better sense of phenomena that were already known. One of the serious intellectual and cultural battles of the period concerned challenges to the Aristotelian ideal of intelligibility, and attempts at replacing it. Such battles offered to supplant Aristotle's kinds of physical explanation by mechanical explanations of natural processes that involved tiny particles or atoms, or by mathematical formalisms that were sometimes associated with the name of Aristotle's teacher, Plato. Adherence to the older models of natural philosophical explanation and the categories that they used long remained a viable intellectual option; it simply became increasingly unfashionable.

Explanatory schemes changed, but so too did research practices. Replacing the Aristotelian stress on known phenomena with one focused on novelty also often involved changing the conception of experience as it was used in the making of natural knowledge. Where Aristotle's "experience" spoke of what was known about how the world routinely behaves, the seventeenth century saw increasing recourse to deliberately fabricated experiments that revealed behaviours that had sometimes never been seen before. Experimental investigation relied on the notion that what nature can be made to do, rather than what it usually does by itself, will be especially revealing of its ways. Francis Bacon spoke of experimentation as being a matter of "vexing" nature; perhaps significantly, as a government agent in the closing years of Queen Elizabeth's reign Bacon had valued torture (an extreme form of vexation, to be sure) as a way of forcing information from taciturn suspects.[6] For Aristotelians, by contrast, the philosopher learned to understand nature by observing and contemplating its "ordinary course," not by interfering with that course and thereby corrupting it. Nature was not something to be controlled.

Thereby hangs a major, and signal, difference between the older academic philosophy of nature and the enterprise that emerged from the "Scientific Revolution." It is as well to have an accurate understanding of what is at stake in the use of that term before attempting to explain it, and the theme of operationalism is as effective a summing-up of the wide body of changes as any. It captures the core issues behind the abandonment of much of Aristotelian views of nature: this was not a critique of means, as Bacon himself observed, so much as of ends.

So do birds *know* how to fly? Does a cook *know* what bread is? Bacon would have answered "no" to the first question, and "maybe" to the second. A cook should not be said to "know" about bread in a philosophical sense simply by virtue of being able to make it, any more than a bird "knows" about flight by virtue of being able to fly. But Bacon believed that a philosophical cook, who already possessed true knowledge about bread, would by definition be able to make it well, because a criterion for knowing truly the nature of something is the ability to reproduce it artificially. The proof of the pudding was in the cooking. Hence Bacon's scorn for Aristotelian natural philosophy: it offered explanations that did not speak to operation; that could not be put to work.

The subject of this book, in short, is a wholesale and profound restructuring of ideas about nature, of the proper purposes of knowledge about nature, and of ways of acquiring that knowledge. The large-scale cultural developments that brought about these new intellectual and social values, and thereby created new senses of what it might mean to understand something, are more than just entertaining window-dressing for a story about the emergence of modern science. They are integral parts of what modern science itself is – what it is about and what its procedures mean.

III Renaissance and revolution

The story can be divided into two stages. Although the term "Scientific Revolution" has long been used for the entire period considered by this book, it will refer here specifically to the seventeenth century. The first of our two stages can, by contrast, be called the "Scientific Renaissance."[7] The period of European history known as the Renaissance is one that, depending on region, lasted from the end of the fourteenth century until the start of the seventeenth. Its relevance to our concerns arises from its broad cultural rôle in most areas of intellectual endeavour, including the scientific, with its widest impact being felt only in the second half of the period. The Renaissance is characterized by a cultural movement that was promoted by people who saw classical antiquity, the world of ancient Greece and Rome, as a model to be emulated in their own time. It spread most effectively through the medium of educational reforms taking place in the schools and universities that trained the élite classes, and its values were therefore widely promoted among the powerful and the learned. From Italy the movement spread northwards across the Alps, transforming cultural life not just among the literate minority but also, through their influence, in society as a whole. For the sciences, it meant above all a focus on the philosophical, including the mathematical, traditions and texts of antiquity. "Renaissance" means "rebirth," and the ancient world that was being revived was one that included, besides the architecture of Athens and the poems of Ovid, the physics of Aristotle, the mathematics of Archimedes and the astronomy of Ptolemy. These were not, to be sure, of major concern to most, but among those who took an interest in them, achievement lay in restoring the endeavours of those classical authors. Our first concern, therefore, will be with this "Scientific Renaissance," which will take us from the end of the fifteenth century through to the beginning of the seventeenth.

The second stage can properly be called the "Scientific Revolution" because it is only in the seventeenth century that the dream of improving knowledge of nature by restoring the ways of antiquity began to be replaced by a widespread sense that newly developed knowledge surpassed, rather than merely emulated, ancient achievements. No longer would the way forward be mapped out by recovering what the ancients had supposedly already known.[8] The gradual acceptance of novelty is a notable element of this story. Even by the end of the century so-called *novatores* (innovators) continued to be criticized in some quarters precisely because they were *not* following the lead of ancient authorities: it was regarded by some as being in rather poor taste. Most of the major names of the standard Scientific Revolution indeed worked in the seventeenth rather than the sixteenth century. The only undeniably major figure of the sixteenth century is Nicolaus Copernicus, perhaps along with his fellow astronomer Tycho Brahe. Others, such as Kepler or Galileo, produced their most important work after the start of the new century, as did René Descartes, Christiaan Huygens, and Isaac Newton among many others. It was in the seventeenth century that increasing challenges to the scholastic-Aristotelian orthodoxy in philosophy became sufficiently powerful to displace it from its previously secure position. Aristotelian philosophy continued to be taught in the colleges and universities of Europe, but by 1700 it was hedged about with many qualifications, some of them profound. Institutional inertia, due to the presence of Aristotle's writings on many official curricula, helped its remnants to limp a good way into the eighteenth century, but the world had changed: knowledge of nature increasingly implied knowledge of how natural things *worked* and how they could be *used*.

Chapter One
"What was Worth Knowing"
in 1500

I The universe of the university

In 1500 the universities reigned over European intellectual life. Their organizational structures were closely modelled on the thirteenth-century prototypes from which they derived, and the content of their philosophical instruction generally conformed to the tenets of scholastic Aristotelianism already described. Those tenets engaged with more than just the formal characteristics of explanation, however; they were also tightly entwined with a picture of the structure and make-up of the physical universe.

Aristotelian philosophy spoke of a spherical universe at the centre of which was found the spherical earth. Aristotle's world, rooted in sense-experience, was always addressed to the position of human observers, not to that of some transcendent, godlike being viewing the whole from the outside. Accordingly, the heavens, above our heads, obeyed different regularities from those observed by things around us on the earth's surface. The heavens revolved around the central earth, cyclically generating the periods of time that structured both the calendar and the daily round. The heavens did not fall down; nor did they recede from us. By contrast, on earth we are surrounded by heavy bodies that fall, and light bodies that tend to rise. Thus the characteristic motions found naturally in the terrestrial realm were either towards the centre or away from the centre; those of the heavens, by contrast, took place *around* the centre.

That way of perceiving things was integrated with a theory of matter. How do we know what things are made of? For Aristotle, the answer is that we see how they behave. On the surface of the earth, there are bodies that fall. These bodies therefore have a characteristic property of heaviness. But not all bodies that fall are the same. Solid bodies that fall are said to be composed primarily of the element "earth," while liquid bodies that fall are said to be composed primarily of the element "water," Both move as they do when they are displaced from their proper locations in the universe. The natural place of earth is at the centre of the universe, whereas the natural place of water is around the natural place of earth – which is why the oceans tend to surround the solid earth. Corresponding to the two heavy elements are two light elements, air and fire, which possess, rather than "gravity," the property of "levity." Thus we see that air-bubbles and flames rise. Air occupies the region above that of water, while fire occupies that above air. The four together exhaust the number of elements making up terrestrial matter.

This terrestrial onion, earth at the centre, water, air and fire in successive shells, occupies only a small proportion of the universe. The vast region beyond the sphere of fire constitutes the heavens, moving cyclically around the centre. Because of that characteristic motion, categorically different from that of the terrestrial elements, the heavens are said to be composed of a single element, the "aether," the natural motion of which is precisely this circular rotation. Indeed, it is on the basis of this routinely observed motion that the existence of aether is inferred to begin with.

The visible celestial bodies, consisting of the moon, sun and five planets (those visible to the naked eye) are carried around the earth by transparent, invisible spheres. These spheres continue the onion motif: they are nested one within the next around the centre, each celestial body being embedded in the side of a distinct sphere. The spheres revolve, carrying the visible bodies around. The stars are out beyond Saturn, the furthermost planet, on the surface of an enormous sphere. The point of the arrangement, again, is to account for what we, inhabitants of the earth's surface, see. The invisible celestial spheres must be there, because the visible celestial bodies have to be moved somehow. Experience-based knowledge, for Aristotelians, is not just a matter of what can be sensed directly, but also a matter of what can be *inferred* from experience.

There were further, more consequential aspects of the heavens that flowed from these considerations. Elements, as we have seen, are characterized by their natural tendencies towards motion, whether up, down, or around. But they could also change into one another, because that is a commonly experienced behaviour: liquids become solids, solids burn to produce fire, and so forth. Part of the concept itself, therefore, implied the possibility of change – at least as far as the terrestrial elements were concerned. The heavens, however, were immune from this kind of transmutation. They were composed of a single element, the aether, a point that necessarily precluded substantial change. Things made of aether could be denser or rarer, but there was no other celestial element for them to change into. Nothing in the heavens came into being, or ceased to exist; celestial motion itself was cyclical, and no genuine novelty had ever been observed beyond the confines of the terrestrial region. Such ephemera in the skies as comets were accounted, almost by definition, as terrestrial. Aristotle held comets to be meteorological phenomena in the upper atmosphere, below

Figure 1.1 *The Aristotelian universe in the sixteenth century, from Petrus Apianus,* Cosmographia *(1539). The order of the planets in distance from the earth is that due to the astronomer Ptolemy, which differs slightly from that of Aristotle himself.*

the lowest sphere which carries the moon around the earth. Terrestrial and celestial were distinct regions, therefore, governed by different physical constituents and correspondingly different physical behaviours. Terrestrial and celestial physics were both part of natural philosophy, but they were different domains.

This was the world promulgated by the university arts curriculum; the world seen, contemplated, and explained by the scholastic natural philosopher. It was a complex universe, but it was also finite in at least two senses. Not only was it of limited spatial dimensions – a huge but bounded globe enclosing all of Creation – but the *kinds* of things that it contained, and the ways in which it behaved, were also strictly limited. Aristotelian natural philosophy specified the categories of things contained in the world, and exhaustively catalogued the ways in which they could be understood. The reason for the absence of innovation and discovery as a significant part of this worldview is that there was no real sense of the natural world as a vast field to be explored; there was nothing genuinely and fundamentally new to be found in it.

It is therefore of relevance to consider that in 1500, at the start of our period, Christopher Columbus's first voyage was only eight years in the past and the Americas had not yet received their name. The availability of geographical discovery-metaphors was much greater in the sixteenth and seventeenth centuries than had been the case previously: Europeans were looking outwards on a world that no longer corresponded to the classical geography found in the much-reprinted standard ancient text on the subject, Ptolemy's *Geography*. The new sense that the world was large, and largely unknown, was not, therefore, purely philosophical.

The sharply defined quality of Aristotle's physics, which provided such a preordained field for natural philosophy, arose from the four causes into which he analysed the categories of human explanation. His basic question amounted to asking "How do we understand things?" His answer was that we, as a matter of fact, understand or explain things according to four models, designated "causes." Together, the four causes are intended to exhaust all the possible ways in which people explain or understand. "Final cause" explanations make sense of the behaviour or properties of something by invoking its purpose: I walk because I'm going towards a destination; a sapling grows because it strives to be a fully-grown tree. The "final cause" is "that for the sake of which" something occurs, in the case of events or processes, or is the way it is – such as explaining the arrangement of teeth in the mouth by reference to their chewing function (this second kind is called "immanent teleology"). The "material cause" adduces what a thing is made of: my chair burns when ignited because it is made of wood, an inflammable material. The "efficient cause" (sometimes called the "moving cause") is closest to our modern understanding of that word: it is the action by which something is done or brought about. Thus the efficient cause of a gun firing might be the pulling of the trigger; or of a snooker ball rolling into a pocket, the preceding collision between it and the cue-ball.

The trickiest, and at the same time most characteristic, of Aristotle's four causes is the "formal cause." This accounts for the kind of explanation that makes reference to the nature of the thing in question. Consider again this classic medieval syllogism:

All men are mortal
Socrates is a man
Therefore Socrates is mortal.

The formal cause of Socrates' mortality is the fact that he is a man – that is the *kind* of thing that he is – and it is in the nature of men to be mortal. The reason for this kind of "cause" being called "formal" is that Aristotelians referred to the kind of thing that something is as its "form."

The concept of forms is central to Aristotelian thought. It arose from a reinterpretation of a general philosophical problem considered by Aristotle's teacher, Plato. How does one recognize what an individual thing is? How does one know, for example, that this tree *is* a tree rather than a bush, or even a helicopter? Plato's answer, in which he was followed by Aristotle, was to say that one must *already know* what a tree is in order to recognize one. And what one already knows, namely what a tree is in general (that is, what *sort* of thing a tree is), Plato describes as knowledge of a tree's *form*. Forms, for both Plato and Aristotle, are in effect categories into which individual objects can be sorted. The category into which something fits (tree, bush, helicopter) represents what kind of thing that object is: Socrates, in the earlier example, is a man. Thus the world is seen as being made up of categories, or classificatory boxes, that take account of everything that exists or could exist. Aristotle's is a vision of the world that sees it as a taxonomic system, in which there is a place for everything. True philosophical knowledge of the world amounts to locating everything in its place. Furthermore, causal properties are important parts of properly categorizing things.

The purpose of this philosophical scheme, therefore, was to understand in the most fundamental way what things *were* and why they *behaved* as they did. And Aristotle's taxonomy of causes determined, as taxonomies tend to do, what could and could not be said of natural phenomena, and what was *worth* saying. At the same time, it should be remembered that, to a greater or lesser extent, this is a property of any classification system, and by extension any framework within which to locate knowledge of nature. It is not the case that Aristotelian philosophy restricted the sciences of nature whereas its replacements extended their scope. Any single system would have had these same structural characteristics, some of which we will see in later chapters. But the abandonment of scholastic Aristotelianism, especially during the seventeenth century, was accompanied by a proliferation of alternatives which, collectively, greatly expanded the possibilities – even if most of those possibilities, different in each case, would have been rejected from within any particular philosophical scheme. In the case of those new systems which were presented in the seventeenth century as something new, there was in addition the prospect of unpacking their implications and following their precepts for the first time, in contrast to the well-surveyed territory of their chief predecessor.

II Natural knowledge and natural philosophy

The scholastic Aristotelianism prevalent in Europe at the start of the six-teenth century differed in some significant respects from the philosophy found in the writings of Aristotle. That philosophy, and particularly its natural philosophical components, had first been assimilated into the aca-demic world of Roman Catholic (or Latin) Europe in the twelfth and thir-teenth centuries. The assimilation wrought its own changes, which sprang from the settings in which Aristotle was seen as being of interest to begin with. Scholars tended overwhelmingly to be clerics, since they were the ones who were much the most likely to be literate. The Church, as the dominant institution throughout Western and Central Europe, played a major role in determining intellectual priorities: Aristotle came to be inter-esting because he could be used to illuminate matters of theological inter-est. After conflicts and disagreements during the thirteenth century, especially at the University of Paris, the works of Aristotle on a whole range of subjects from logic and rhetoric to meteorology were securely ensconced in the curricula of the new universities, even while official Church dogma still tended to circumscribe some aspects of their interpretation. The theo-logical value of natural philosophy stemmed straightforwardly from its topical focus: interpreted from a Christian standpoint, it concerned God's Creation. Learning about God by learning about what He had made, and understanding the whys and wherefores of its fabric, was seen by many as an eminently pious enterprise. Natural philosophy had become a religious endeavour, and it remained so for many centuries. In the early eighteenth century, Isaac Newton wrote that "to treat of God from phenomena is cer-tainly a part of natural philosophy."[1]

This is not to say, however, that natural philosophy in medieval and early-modern Europe was *always* understood as dealing explicitly with the natural world as God's Creation. Usually it was, but, as Newton's pro-testation suggests, the connection was not a necessary one. In sixteenth-century Padua (a leading university centre), as also in thirteenth-century Paris, so-called "Averroïsm" caused great controversy by purporting to discuss Aristotelian natural philosophy in isolation from a Christian theo-logical context. The twelfth-century Arab philosopher Averroës had written extensive commentaries on Aristotle's natural philosophical writings that attempted to explicate their content independent of extraneous reli-gious doctrines (in Averroës' case, Islamic). In the thirteenth century some Christian scholars at Paris followed Averroës' lead, developing his inter-pretations of Aristotle in sometimes flagrant disregard for theological controversy. Their frequently condemned attempts to get away with this relied on the possibility of representing their endeavour as being natural philosophy and *not* theology. Natural philosophy was clearly not invariably seen as a study of the divine. Their position was, however, opposed by such alternatives as Thomas Aquinas's. Aquinas made an extremely influential attempt in the thirteenth century explicitly to disallow Averroïsm; his view of natural philosophy as a "handmaiden" to theology quickly became commonly accepted, and coloured the conception of the discipline thereafter. In practice if not always in principle, natural philosophy and theology had become inextricably linked.

The university world of 1500 had expanded significantly since the foundation of the first such institutions around 1200. The word "university" is an English version of the Latin *universitas*, a term routinely applied in the medieval period to legal corporations. Only over the course of centuries did "university" come to be associated specifically with those corporations (whether of scholars or of students) devoted to educational purposes and offering various grades, or "degrees," through which the student attempted to pass. The fifteenth century saw a rapid increase in the number of universities across Europe, largely due to the foundation of new institutions in the eastern parts of Catholic Europe, such as Poland (Nicolaus Copernicus studied at Krakow in the 1490s). The new foundations retained the same basic organizational structure as their medieval prototypes, however. Their basic component was the so-called Arts faculty, the division that dealt with the "liberal arts" of which philosophy (natural, metaphysical, and moral) was the major component. Following successful passage through the degrees of Bachelor and then Master of Arts, students aiming at a doctorate in a professional discipline went on to study in one of the three "higher" faculties of medicine, law, and theology. In the non-Italian universities, north of the Alps, theology was usually the most important of the three. This vocational direction tended to affect the treatment accorded to natural philosophy, reinforcing its perceived role as a handmaiden to theology.

A characteristic shared by all three of the higher faculties, however, and not just theology, was that they served vocational directions that were not open to women. It is therefore unsurprising that there was virtually no place for women in the universities; the basic purpose of the university was to train young men in one of the professions. The most characteristic, and important, vocation in the Middle Ages lay in the church – perhaps the exemplar of a major social institution restricted to men only. Clerics could in principle, and to varying extents did, come from all social classes; but they could never be female. This fact is probably too deeply rooted for its implications to be easily and unequivocally traced, but it has been suggested that the longstanding domination of western science by men may owe something to the clerical character of its academic and scholarly origins. What effect that may have had, in turn, on the conceptual and ideological structure of the sciences cannot be clearly stated, owing to the vast number of mediations that would have to be traced to make the relevant connections. Nonetheless, it will be important to bear in mind this basic sociological fact about the knowledge enterprises of medieval and early modern Europe in what follows.

Besides natural philosophy itself, there were other subjects of study concerning knowledge of the natural world that were also taught in the universities. Medicine, one of the higher faculties, involved study of such components as anatomy and *materia medica*. The anatomy of the human body was increasingly, by 1500, being taught to medical students at northern Italian universities and elsewhere in part through demonstration-dissections, whereby a corpse would be dissected over the course of several days for the benefit of onlookers. The textual accompaniment to these displays was typically a digest of the anatomical teachings of the ancient physician Galen (late second century AD), such as the early-fourteenth-century handbook by the Italian Mondino de' Liuzzi. The purpose, however, was not to conduct research; it was wholly pedagogical, intended to familiarize students with the internal structure of the human body according to Galenic doctrine. The area of *materia medica* concerned such things as drugs and ointments, together with their preparation from mineral and especially botanical sources. It therefore included natural historical knowledge of plants and their medicinal properties. It might be noted that neither of these studies, anatomy or *materia medica*, purported to deal centrally in philosophical content. Although the human body and its parts were to be understood in terms of Galen's theoretical (really, natural philosophical) views, the study of anatomy itself concerned detailed morphological description rather than being focused on explanation. *Materia medica* was also a field that presented practical know-how rather than theoretical understanding. When dealing with plants, for example, the physician was not concerned with the causal science (in Aristotle's sense) of botany, but with empirical knowledge of a plant's properties. While the language of natural philosophy was often used to characterize the medicinal properties of a drug, it was an auxiliary aid to medical knowledge rather than an end in itself.

The other main area of natural knowledge that was separate from natural philosophy, this time in principle as well as in practice, was that of mathematics. The chief mathematical sciences practised in the medieval university were astronomy and, to a lesser extent, music. These were both members of the medieval *quadrivium*, comprising the four mathematical sciences of arithmetic, geometry, astronomy and music. The theoretical justification of this grouping conformed once again to Aristotelian expectations: the first two were the branches of "pure" mathematics, dealing with abstract magnitude as their proper subject matter. Arithmetic was concerned with discontinuous magnitude – numbers – whereas geometry concerned continuous magnitude, in the form of spatial extension. The third and fourth members of the quadrivium represented the branches of "mixed" mathematics. That term signalled that their proper concern was magnitude *combined with* some specific subject matter. Thus astronomy was geometrical extension as specifically applied to the motions of the heavens; music was numbers as specifically applied to sounds.

These last two disciplines were thus sciences of the natural world, but were explicitly denied the status of natural philosophy. Aristotle himself had characterized such subjects as mathematical, distinguished from natural philosophy by their supposed lack of causal explanations. The mathematical astronomer, on this view, merely described and modelled the motions of the celestial bodies; it was the job of the natural philosopher to explain *why* they moved. Similarly, the mathematical musician codified the number-ratios corresponding to particular musical intervals (number-ratios that found their typical physical instantiation in musical string lengths). The natural philosopher was left with the task of explaining the underlying *nature* of sound.

This Aristotelian characterization of the mathematical sciences informed the curricular structure of the European universities of 1500. Natural philosophy was taught as an important component of an arts education, while mathematical studies, when they were given any significant place, tended to be presented as independent, specialized disciplines that were chiefly aimed at practical ends involving computations of various sorts. Astronomy was the most important such science in the universities, for a number of reasons. First, its practical functions were highly regarded: these included calendrical uses, such as computing the dates of moveable church feasts (although this had become routine and unproblematic by the time of the foundation of the first universities), and the casting of horoscopes. Astrology was not a specialty definitively distinguished from astronomy; the astronomer was also an astrologer, while the astrologer always had to have command of astronomy, the science of the motions of the heavens. The high practical importance of astrology stemmed from its use in learned medicine, where the casting of horoscopes was a routine procedure in making prognoses regarding the future course of an illness. Indeed, astronomy was a particularly prominent study at the University of Padua during the Middle Ages because the faculty of medicine was the major of the three higher faculties there, a priority that reflected back onto the preparatory arts training.

III Astronomy and cosmology

The relationship between astronomy and cosmology in the medieval university tradition was an uncertain one that only began to change in major ways in the generation or so preceding Copernicus. The word "cosmology" in its modern sense dates from the eighteenth century, but it is useful to apply it to earlier conceptions of the physics of the heavens and to related natural philosophical ideas about the overall structure and workings of the universe. The cosmology found in the university world preceding

Copernicus's work was broadly Aristotelian, as we saw above, and as such it restricted the heavens to motions that were perfect, uniform, and circular – the kind of natural motion appropriate to spheres composed of aether. The foremost astronomical authority of Antiquity, Claudius Ptolemy (second century AD), had written the work that became the bible for subsequent astronomers in the Greek, then Arabic and Latin traditions. It was known in the Middle Ages as the *Almagest*, a corruption of the usual title in Arabic; in Greek it was the *Syntaxis*, or "compilation." Ptolemy had commenced the *Almagest* with a brief treatment of the physical framework that constrained his accounts of the motions seen in the heavens, and it was a framework derived from the natural philosophy of Aristotle. Ptolemy followed in the tradition of previous Greek astronomers in restricting the elementary motions from which actual observed motions were to be synthesized to uniform, circular motion, and that restriction (held to have originated with Plato) itself seemed to conform to Aristotle's account of celestial physics. While the physics to which Ptolemy deferred was Aristotelian, the vast bulk of the *Almagest*, having adopted Aristotle's cosmological ground rules, followed the autonomous mathematical traditions of Greek astronomy. Ptolemaïc astronomy as it was practised in the Latin Middle Ages was therefore easily cordoned off from natural philosophy in the universities.

Not all the complexities of Ptolemy's achievement, let alone the refinements and improvements furnished by several intervening centuries of Arabic astronomy, were adopted by Latin astronomy following the translation of the *Almagest* from the Arabic in the twelfth century. Ptolemaïc astronomy, having justified its central, stationary spherical earth with Aristotelian arguments, arrayed the heavenly bodies in orbits around it, accounting for the details of their paths by using an array of subsidiary circles. A (very) simplified Ptolemaïc model of the motion of a planet would look like Figure 1.2. (This diagram discounts the daily motion of the heavens; strictly speaking, the entire diagram ought also to revolve around the central earth once a day.) A simple circular path around the central earth would not have fitted observation: although, in general, the heavens appear to roll around the earth in circular paths (hence the characteristic properties of Aristotle's elemental aether), the planets show anomalies. Mars, Jupiter, and Saturn, for example, periodically slow up in their overall course from west to east through the stars, and double back away before continuing in their original direction; the doubling-back is called "retrograde" motion, the phenomenon itself "retrogression." The smaller circle in Figure 1.2 allows this appearance to be mimicked. The planet moves uniformly around the small circle (called an "epicycle"), while that circle's centre revolves uniformly around the larger circle (called a "deferent"). If the lesser circular motion completes several revolutions for each single revolution of the greater circle, the appearance from the centre will be of the planet looping back in its motion whenever it passes around that part

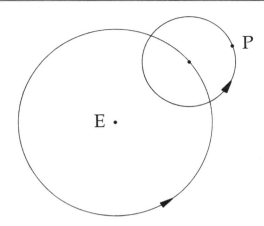

Figure 1.2 *Simplified basic planetary model, as used by Ptolemy. An epicycle, around the centre of which the planet is carried, itself in turn revolves on its deferent circle around the central earth.*

of the small circle which is on the side closest to the system's centre (Figure 1.3). This conception formed the basis of Ptolemy's explanations of planetary motions around the earth. To achieve the greatest possible accuracy, many refinements, including additional subsidiary circles, needed to be made to models of this kind.

From the point of view of a natural philosopher, however, this approach would have been questionable if presented as an *explanation* of planetary motions. Not only is there no attempt to explain *why* such circles move as they do, or what those circles are composed of; the circles themselves (in this case, the epicycle) routinely have as their centres of revolution points displaced from the centre of the earth (and hence of the universe). By contrast, Aristotle's concept of the circularity of celestial motion involved an understanding of that motion as being centred on the earth. What, then, was the physical status of Ptolemaïc models in the Middle Ages?

From one perspective, they were simply calculating devices. As long as the numbers that they generated corresponded to observed celestial positions, it mattered little to the mathematical astronomer whether the details of the models represented real motions in the heavens or were fictitious. Were epicycles and deferent circles real objects, or figments of the astronomer's imagination? From a practical standpoint, the astronomer did not need to worry about it; a resolution of the question would not improve his calculations. But for the natural philosopher, things were less simple. While the astronomer ought not to worry about physical causes, such questions were the natural philosopher's stock-in-trade.

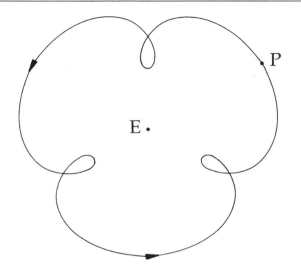

Figure 1.3 *The epicycle in the Ptolemaïc planetary model accounts for the periodic appearance of retrograde motion as seen from the earth, during the times at which the planet is at its closest approach.*

Usually, however, medieval cosmology concerned itself with general questions of the nature of the heavens and the causes of celestial motions, leaving the details of those motions to the astronomers as if it made little difference. Only very seldom did medieval natural philosophers so much as consider questions concerning the physical status of the astronomer's complex systems of circles. It was easy to ignore them, evidently, in light of the hierarchy of the disciplines in the universities: natural philosophy, dealing with causes and the natures of things, was more highly regarded than the more practical craft of astronomical computation.

Physicists were in a position, therefore, largely to disregard the concepts used by astronomers, much as a botanist might ignore the practical wisdom of the gardener. For their part, astronomers ignored the same issues of the compatibility between the physics and the mathematics of celestial motions even more completely than did the physicists; astronomical treatises of the Middle Ages do not broach the subject at all.

This was the situation from the introduction of the *Almagest* into Latin Christendom until the second half of the fifteenth century – the generation before Copernicus. Since the thirteenth century, one of the major teaching texts on astronomy had been an anonymous work called *Theorica planetarum*, or *Theoric of the Planets*. The title's first word, usually translated into English as "theories," refers rather more specifically to the particular geometrical models of the motions of celestial bodies that the book contains; *theorica*, or "theoric", here means something like "theoretical modelling." The work contains such models for the sun, moon, and the five planets, together with instructions on how to use them for computation. It is important to note that, although clearly derived from the models given in Ptolemy's *Almagest*, those of the *Theorica* are distinctly simpler, eschewing much of the complexity that Ptolemy had used to achieve a high level of accordance with observational data. In this connection, it is also relevant to note that the *Alfonsine Tables* (*c.*1272), the standard numerical tables used in the Middle Ages for determining celestial positions, had been computed for each of the celestial bodies from geometrical models that were enormously simplified compared with their Ptolemaïc prototypes. As long as the predictions that could be made were good enough, precision for its own sake was not a desideratum – another indication of the practical bent of medieval astronomy.[2]

In the 1450s, however, a German astronomer in Vienna, Georg Peurbach (or "Peuerbach"), wrote a new teaching text bearing the title *Theoricae novae planetarum*, or *New Theorics of the Planets*, finally printed in 1475. As the title suggests, it was intended as an improved replacement for the old *Theorica planetarum*. It presented the same kind of material as its predecessor, endeavouring only to improve certain features of the individual models but in no way attempting to present models of the same complexity as those found in the *Almagest*. Perhaps the most radical innovation, however, lay in its presentation of those models themselves. Rather than showing diagrams made up of geometrical lines representing distinct motions, as in Figure 1.2, above, Peurbach displayed solid spheres of a finite thickness (Figure 1.4).

The sun moving on a deferent has turned into a body embedded within the walls of a deferent sphere which itself is embedded within a much larger, hollow sphere that encompasses the earth. This unquestionably *physical* picture was much more compatible with the physical spheres spoken of by Aristotle and the scholastic natural philosophers than was an abstract, computational geometrical model. It is the first time that an astronomer, rather than natural philosopher, in the world of Latin Christendom had confronted the issue of the physical status of his models: Peurbach wished to interpret his mathematical devices as having physical referents. Observational niceties had led mathematical astronomers to add circle upon circle to the basic Aristotelian picture of the heavens; now Peurbach insisted on regarding those circles as physical things that astronomy had, in effect, discovered – *discovered*, because these were circles that apparently needed to exist in order for the appearances in the heavens to be as they were. He declined, that is, to regard mathematical astronomy as purely instrumental, like most navigational calculations today, which are still conducted, for convenience, according to the fiction of a stationary earth because they give the right answers.

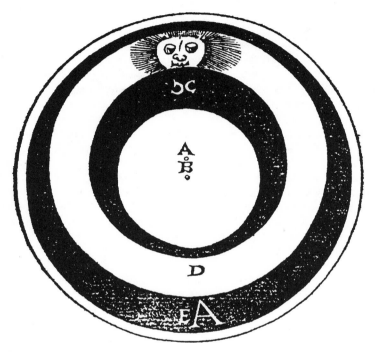

THEORICA

THEORICA TRIVM
orbium Solis.

Figure 1.4 *Peurbach's conception of the physical reality of Ptolemaïc astronomical models, from his* Theoricae novae planetarum: *each circle in the geometrical model is here interpreted as a three-dimensional solid in the heavens. B represents the central earth; A is the point (axis, normal to page) around which the eccentric sphere D, which carries the sun, rotates. The planets are handled using similar, but more complicated, techniques.*

Ptolemy's astronomy reached Copernicus, at the close of the fifteenth century, through accounts such as Peurbach's; that is, accounts at one remove from the *Almagest* itself. His most detailed source was a work by Peurbach and his collaborator Regiomontanus, called *Epitome of the Almagest.* Although completed in the early 1460s, this work was not printed until 1496 (in Venice), by which time Copernicus had begun his university studies of astronomy. The *Epitome* is what its name suggests: a digest of the *Almagest* intended to illuminate the niceties of Ptolemy's geometrical models for celestial motions. Much more elaborate than either the *Theorica planetarum* or even Peurbach's *Theoricae novae*, the *Epitome of the Almagest* represented the peak of Latin astronomical science in 1500. But it was an astronomical science that already threatened encroachment on the domain of causal natural philosophy by implying its competence to speak of the true workings of the heavens.

IV Beyond the university

A scholar's life by 1500 was as much material as intellectual. The end of the fifteenth century saw a European learned culture that was busily absorbing the impact of a new technology, that of printing with moveable type. First appearing around 1450 in the German city of Mainz, printing rapidly spread from Johann Gutenberg's original press throughout the German territories and northern Italy, most notably Venice. This establishment, during the second half of the century, of scores of print shops corresponds to two related features of European, especially Western European, society at that time. The first is the fairly high rate of literacy on which the market for books and pamphlets was based. The second is the quite sudden wide availability of a multitude of philosophical and general intellectual options. Together, these two features created a situation in which knowledge for very many people was no longer so chained to the texts of the university curriculum. This was a new situation practically without parallel.

In 1500 the variety of intellectual options being sought in the new literary environment was still relatively limited. Perhaps the most influential among them at the time was the so-called neo-Platonism of Marsilio Ficino in Florence. In the 1460s Ficino undertook to translate into Latin the works of Plato, which had received relatively little attention during the Middle Ages.[3] Prior to this task, he had translated in addition a number of texts nowadays reckoned to date from the early Christian era, but which were believed by Ficino and most others to be among the oldest texts of antiquity, predating by centuries the writings of Aristotle, Plato, and all the other luminaries of classical culture. These were texts of the so-called hermetic corpus, held to have been written by an ancient sage in Egypt called Hermes Trismegistus (thrice-great Hermes). Their most notable feature (apart from reports of ancient Egyptian temple magic, in a hermetic text already known to the Latin Middle Ages) was their metaphysical conception of the universe. This, not surprisingly, resembled closely the neo-Platonic doctrines of late antiquity, the period now reliably believed to have produced the "hermetic" corpus as well. Since the late-antique neo-Platonic writers, such as Plotinus (third century), regarded their own teachings as expositions of the more arcane and mystical implications of Plato's own work, Ficino believed that all three together – Plato, the hermetic corpus, and the neo-Platonic texts – represented an ancient mystical tradition of profound wisdom dating back to before the time of Moses in the Old Testament.

The overall thrust of these doctrines was a picture of the universe as a spiritual unity, in which the various parts were related by spiritual sympathies and antipathies. Astrology was a characteristic aspect of these views, one that was widely shared. Astrology had, of course, been a standard part of the learned beliefs of the preceding centuries too. Its novel aspect in the new neo-Platonic or hermetic form championed by Ficino, however, involved the ambition to use the astrological influences of the stars for human ends, rather than simply to predict and document passively their effects. In other words, this was a kind of astrology that held out the dream of magical domination of nature. The astrologer-magus – of whom another good example from the late fifteenth century is the Florentine Giovanni Pico della Mirandola – dreamed of harnessing the powers of the stars in their psychical interaction with things on the earth.

It can be noted right away that this neo-Platonic magical strain in Renaissance thought, clearly distinct from the usual teachings found in the university, does not strictly count as a non-Aristotelian *natural philosophy*. This is because it endeavoured to be more than that. To the extent that its view of nature was directed towards achieving operational control over nature, it was centrally a form of *magic*, a kind of technology intended for practical ends, and not a philosophical study devoted to understanding for its own sake. However, at the same time, because it did implicate non-Aristotelian philosophical ideas about the workings of the universe, it was one route (among many) by which non-Aristotelian natural philosophies became established in competition with orthodoxy. Magic was certainly known in the Middle Ages too, but it was usually presented as, at the least, not incompatible with Aristotelian natural philosophy. The Aristotelian world was a world of regularities, but not a world of rigid determinism. Unusual things could sometimes happen, and magic attempted to operate in that rather lawless hinterland left out of account by Aristotle's emphasis on what *usually* happened.

Magic itself could be a treacherous category. At its most fundamental level, the term referred to an art of manipulation, of doing things that, specifically, tended to provoke wonder, or that were marvellously out-of-the-ordinary. There was, in consequence, a variety of practices that properly bore that label. They included spiritual magic, which worked by invoking the aid of spirits, whether angels or demons (the latter then being known as "demonic" magic, associated with witchcraft); and "natural" magic. The latter was supposed to work by exploiting, rather than the abilities of spiritual agencies, the hidden ("occult") powers found in nature. The action of a magnet upon iron, for example, manifested one fairly common such power. Magic was an art of doing things, a technology, and the magus was someone who knew how to use it. It therefore represented a quite different kind of knowledge from the Aristotelian contemplative ideal.

Operational knowledge manifested itself in other ways too. The advent of printing saw the appearance not only of Latin treatises on magic, accessible to the learned, but also texts in the vernaculars, such as dialects of Italian and German. These vernacular texts had, of course, a much wider potential readership. Literacy was required, but not the scholarly training that would render Latin texts accessible. The new "books of secrets" therefore presented practical, but usually rather recondite, information to people with only a middling education. The genre, which really took off in the sixteenth century, seems to have been very much a creature of printing; the demand for such books was fed and encouraged by printers who could promise their readers all manner of practical tips that had hitherto (or so it was claimed) been the preserve of closed guilds of practitioners. Medical advice was particularly popular, with books presenting recipes for the treatment of a wide variety of ailments. In the first half of the sixteenth century one of the most prominent authors of such vernacular medical texts was Walther Hermann Ryff, a man with some training in medicine and the arts of the apothecary. Ryff published a multitude of popular works in German, largely drawn from the writings of others in the same fields, and including material taken from the learned Latin treatises of the university medical schools. In 1531 and 1532 there first appeared a group of small booklets, known as *Kunstbüchlein* ("little craft-books"), on a variety of practical craft and technical subjects. These anonymous books were produced from the shops of printers in a number of German cities, and catered to what they revealed as an eager appetite for such things not just among German craftsmen, but among literate people of the middling sort in general. They broke the perceived monopoly of the craft guilds over possession of such practical knowledge as made up metallurgy, dyeing or other chemical recipes, pottery or any of a multitude of potential household requisites.

The historian William Eamon, in his studies of such literature, has characterized these "technical recipe books" as a means whereby the "veil of mystery" that had hitherto surrounded the practical crafts was lifted, so that ordinary people could see that the craftsman was not possessed of some arcane wisdom, but simply had knowledge of a set of techniques that, in principle, anyone could apply.[4] This is not a notion that should be taken for granted, however. Studies in recent decades of the ways in which expert knowledge is constituted and passed on suggest that practitioners do indeed possess skills that are communicated only with difficulty. Their practical knowledge is often unlearnable from the eviscerated accounts that appear in the pages of experimental papers (in the sciences) or technical manuals (in skilled craftwork in general).[5] Thus, if Eamon is right, the growing sense that developed during the sixteenth century, as a consequence of printing and its uses, that practical craft knowledge ("know-how") can be reduced to straightforward rules of procedure that can be acquired readily from books, was to a large degree an illusion. If this is so, it is an illusion that we have inherited.

Two additional items to the emerging cultural mix deserve mention. Alchemy was to gain more adherents as time went on, until we find Isaac Newton, towards the end of the seventeenth century, as one of its principal exponents – at least judging by the amount of surviving manuscript material. Alchemy, as the name suggests, had been known to the Middle Ages originally from Arabic sources. By the start of the sixteenth century, it had appeared in some printed discussions, albeit generally in an equivocal way. One of the hallmarks of alchemy was its secrecy; writings on the subject were intentionally allusive and obscure, since this arcane knowledge was not to be made available to everyone. Only those who were already in the know were supposed to be able to benefit from texts on the subject. However, so-called alchemy did sometimes appear in the new printed genre of "books of secrets." This kind of alchemy differed, however, from the mystical alchemy practised by magicians, the kind of alchemy that had close ties to astrology.[6]

Thus the first of the *Kunstbüchlein*, appearing in 1531, was entitled *Rechter Gebrauch d'Alchimei* ("The Proper Use of Alchemy"). Based on a genuinely alchemical treatise that concerned itself with such matters as transmutation, this printer's compilation restricted the contents largely to practical metallurgical and chemical techniques; a kind of workshop *vade mecum*. Clearly, from this perspective, the "proper use" of alchemy was one that divested it of its more speculative and mystical aspects.[7] Despite this, right through to the time of Newton alchemy remained closely identified with spiritual and mystical dimensions. Thus, famously, one of the factors supposedly affecting the outcome of an alchemical preparation was the spiritual state of the alchemist; failure in such work did not necessarily reflect upon the techniques used, but might simply mean that the alchemist's soul had not been sufficiently pure. A transmutation could only be effected if the alchemist's spiritual rapport with the materials being manipulated was of the correct kind. Alchemy was usually by its nature a secretive practice, rather than a publicly available set of techniques suitable for publication in a handbook; witness the radical editing performed for the 1531 text.

Another secretive and magical domain of knowledge around this time was cabalism. Adverting, like neo-Platonism, to the clandestine knowledge of late antiquity, cabalism had originally been a Jewish practice, of which the Renaissance saw the emergence of a Christianized version. It rested on the investing of words, typically names, with occult significances and interrelationships based on the letters (in the original form, Hebrew) by which they were written. A word could be assigned a numerical value given by the sum of the numbers that conventionally corresponded to its individual letters; two words that had the same numerical value were deemed to have some hidden, deep correspondence. Thus the Christian cabala endeavoured to show that the name of Jesus really did correspond to "messiah," by showing that those two words had the same value – a matter in this case of trying to turn Jewish mysticism to the task of convincing Jews themselves of the truth of Christianity. From the 1490s the most prominent writer in this tradition was the German mystic Johannes Reuchlin.

This considerable variety of intellectual options, closely associated with the new technology of printing, meant that Europe around 1500 was preparing itself for a battle over intellectual authority of epic proportions. The sixteenth century was to see one of the great upheavals of European history with the Protestant revolt against the Catholic Church, a rejection of ideas and systems of authority that had held most of the continent for centuries. Placed alongside the questioning of papal authority, challenges to Aristotelian philosophical approaches seem of small significance by comparison. Both, indeed, can be seen as facets of the same process: Martin Luther and Jean Calvin, the most prominent of the religious reformers, both put a stress on the text of the Bible, which was to be made available to all Christians in their own languages, as the cornerstone of the Christian religion. Products of the printing press were to circumvent the elaborate structure of the Catholic Church, to put believers into direct communion with the word of God.

V Learned life and everyday life

It is important to remember the sort of ideas about knowledge of nature that are at issue here. In the case of religion in the sixteenth century, the changes due to the Protestant Reformation and the Catholic response of the Counter-Reformation affected, to a greater or lesser extent, everyone in Latin Christendom. The Counter-Reformation, however, was much more driven by the church hierarchy than was the Reformation, which had involved a great deal of popular religious upheaval in addition to the organized dissent stemming from religious leaders like Luther. The new options in the study of nature in this sense resemble the former more than the latter: the intellectual élite (those who presumed to define "what was worth knowing") fomented or opposed the struggles of the period, with little resonance at the popular level. Indeed, it is unclear how much difference the classic "Scientific Revolution" of the sixteenth and seventeenth centuries made to ordinary people. Its innovations left most features of their everyday lives unchanged; the changes that occurred are usually attributable to identifiably different causal factors, such as religious beliefs and practices themselves.

A longstanding view of the classic Scientific Revolution has emphasized the "decline of magic" by the end of the seventeenth century.[8] This view held that belief in witchcraft and other magical, supposedly "irrational" components of the European world-picture crumbled in the face of advancing scientific rationalism. It is, however, a view that carries much less credibility in light of the historical researches of recent years. The popular credibility of such things as witchcraft and astrology remained strong well into the eighteenth century, and there was a significant level of belief in them even on the part of scholars right through to, and beyond, the end of the seventeenth century. The traditional notion of rising "scientific" attitudes sweeping away the relics of superstition no longer seems very satisfactory. An indication of this point may be had from consideration, once again, of the claims made by those eighteenth-century figures who were the first to characterize the preceding couple of centuries as having seen a philosophical "revolution." The motivations driving so noisy a set of protestations stemmed from a desire to defeat utterly those institutions and ways of thought that many such eighteenth-century thinkers opposed. The most visible and powerful upholder of the supernatural was, of course, the Church (in France, the established Catholic Church). With its miracles, demons, and angels, this was therefore the main target of the "rational" philosophers of the new century: if there had not in the eighteenth century still been widespread belief in such things, there would have been no need to proclaim their outdatedness with such ferocity.

It is worth bearing this in mind from the outset, because the picture of a superstitious and credulous Europe in 1500 giving way, by 1700, to a cool, rationalistic, scientific Europe continues to have a strong hold on our views of the past. The astrology, demonology, and so forth of fifteenth-century figures like Ficino were ingredients of the intellectual ferment of the next couple of centuries; they were not photographic negatives of a new rationality that would sweep them away. History is seldom so neat.

Chapter Two
Humanism and Ancient Wisdom: How to Learn Things in the Sixteenth Century

I Language and wisdom

The new challenges to scholastic philosophical orthodoxy in the universities appeared from what seems at first an unlikely source. One of the usually unstressed aspects of an arts education in the medieval university had been the teaching of the subjects comprising what the early Middle Ages (*c*.600 AD onwards) had dubbed the "trivium." The three parts of the trivium consisted of grammar, logic, and rhetoric. The three went along with the so-called "quadrivium" – the mathematical subjects of geometry, arithmetic, astronomy and music – to make up the "seven liberal arts." These had been the basis of higher education in classical antiquity, and their echoes (with the new names "trivium" and "quadrivium") informed the educational norms of the early medieval period in the west.

The seven liberal arts only loosely structured the curriculum in the new universities of the thirteenth century. Logic, with its newly-available Aristotelian texts rather than just early-medieval digests, blossomed in importance alongside natural philosophy and metaphysics in the university arts curriculum. The quadrivium, meanwhile, enjoyed varying fortunes at different periods and among different institutions, but was never (with the very partial exception of astronomy, as we have seen) strongly emphasized. The other subjects of the trivium, grammar and rhetoric, received similarly short shrift. Latin grammar had evidently become the province of pre-university education (it was in effect a prerequisite to university study, since all instruction took place in Latin). Rhetoric, a discipline concerned with modes of persuasion, occupied a minor place, since the study of argumentation was regarded almost exclusively as the province of logic. But the academic status of rhetoric came to change radically in the fifteenth century.

The learned culture that underpinned the period of the Renaissance (dating from around 1400 or so onwards) is usually designated by the term "humanism." "Humanism" is a much later historians' term derived from the contemporary Latin expression *studia humanitatis*. In Italian universities in the fifteenth century, the *studia humanitatis* were those disciplines concerned with language usage – grammar, rhetoric, and poetics. They placed at their core correct Latin (in time, Greek was added), and elegant literary style in composition. The expansion of these studies at a number of the universities of northern Italy went along with an increasing self-assertion on the part of those who taught them. These teachers, the original "humanists," claimed with increasing volubility the importance of their subjects in the arts curriculum as against the scholastic Aristotelian philosophy that had hitherto dominated it.

The local conditions of northern Italy played a considerable part in bringing this situation about and in fostering its success. The entire Italian peninsula was a patchwork of small states, the typical model in the north being the city-state, such as Milan, Venice, or Florence. Each city, with its surrounding territory (often, as in the case of Florence, considerable and enveloping a number of other major towns), thus exercised a high degree of political autonomy, and civic life within them often involved the participation of their leading citizens rather than being subject to the power of a prince. The early humanists took advantage of this situation by stressing the value of a humanistic education to the creation of an active, politically responsible citizen. A training in the *studia humanitatis*, they proclaimed, was a much better preparation for the future citizen than the dry logic-chopping offered by the Aristotelian philosophers. The humanists taught, as the real pay-off of the education that they offered, skill in rhetoric that would serve well the budding political orator, not just in teaching him tricks of delivery but in developing within him the wisdom and judgement required of a statesman.[1] The great model for such a person was the ancient Roman orator and senator from the first century BC, Cicero.

To the humanist educator, Cicero embodied all the virtues of the good republican statesman. During the latter days of the Roman republic, Cicero had been a political leader whose speeches to the senate, regarded by all as classics of effective oratory, still survived to be studied. Furthermore, Cicero had written on the art of oratory, laying down advice and rules on how to compose and deliver a successful public address. Thus both the theory and practice of rhetoric could be found in Cicero's writings, which to the Renaissance humanists were a treasure-trove to be exploited in the present day. Cicero, in short, could show modern society how public life ought to be conducted. The essential trick to be accomplished here, of course, was bridging the gap between effective speaking and the sound policy judgements that political oratory should, ideally, deliver. Again, Cicero was a convenient model. Not only had Cicero spoken and written well; his speeches were also admired for the wisdom of their content. The new humanist ideology contended that the two were, in fact, inseparable: one could only match Cicero's eloquence if one had acquired his wisdom, since only true wisdom could give rise to such eloquence. The art of Ciceronian oratory thereby acquired an almost mystical quality, and by equating the medium with the message, humanist educators attributed to the education that they provided a privileged role in creating good citizens. In practice, this meant that pupils were trained to *imitate* Cicero's Latin style, as well as imitating the styles of other good classical authors. An eloquent fool was not to be entertained; it would be an oxymoron, or so humanist educators maintained.

The fifteenth century had seen the gradual spread of the humanist educational agenda. It crossed the Alps around the middle of the century, and by the 1490s strongly coloured the style and content of university curricula in countries as far afield as Poland. During the sixteenth century, humanist education created a common cultural style among élite classes everywhere, becoming firmly established in the universities as well as in other kinds of advanced schooling. During the fifteenth century, humanist reformers, most notably Lorenzo Valla, had often attacked scholastic philosophy and theology on both scholarly and moral grounds. They castigated the language of the scholastics for deviating from classical norms – medieval scholastic Latin having drifted away from Ciceronian standards in both vocabulary and grammar – and argued that this barbarism was compounded in the case of theology (especially the "Thomist" theology of Thomas Aquinas). There, bad Latin was deployed in the service of Aristotelian logical niceties regarding metaphysics, to yield a form of theology that seemed far removed from the simple faith represented in the New Testament.

The picture that emerges in the sixteenth century, by contrast, is one of coexistence rather than conflict. In effect, the humanists won their battle for recognition without vanquishing their erstwhile rivals, the scholastic philosophers. Instead, the values of humanism pervaded scholarship as a whole, drifting up from the renewed emphasis on the "trivial" values of rhetoric and classical literature, the revivified remnants of the old trivium. Philosophers had by now routinely received humanistic training under the new educational dispensation. As an almost inevitable result, one finds in the sixteenth century scholastic commentaries on the works of Aristotle that are written in humanistic, classical Latin instead of the barbarous Latin of the medieval scholastics, and consider the niceties of the original Greek texts rather than concentrating solely on medieval Latin translations (themselves "barbaric").

There was much more to the new humanistic scholarly ethos of the sixteenth-century universities than elegant Latin, however. Cicero was a role model for the humanist rhetorician because he had combined eloquence with wisdom in the conduct of civic affairs. His perceived preeminence rested on the assumption that classical antiquity had seen the highest achievements in all areas of culture, achievements that had not since been equalled, much less surpassed. So the greatest orator-statesman of antiquity was, practically by definition, the greatest orator-statesman of all time. By the same token, the greatest practitioners or authorities in practically all other areas of endeavour themselves served as pre-eminent models. Thus the improvement of present-day cultural and scholarly activities came increasingly to be seen as a matter of restoring the highest accomplishments of the ancients. Not progress, but *renewal* was the humanist watchword. The wisdom of the ancients should be sought, in order to reverse the decline that had been occurring ever since the last days of the Roman empire.

II The scientific renaissance

The word "renaissance" means "rebirth." The humanists liked to characterize their own time by using such terminology, because they pretended to be bringing about a rebirth of classical culture. In doing so, they were rejecting the barbarism of the period that intervened between classical antiquity and its rebirth in the present – the "Middle Ages"; it is to the humanists that we owe that name.

The ideal of renewing culture by a return to antiquity first appears in the sciences with any prominence in the mid-fifteenth century. The central figure here was Regiomontanus, a.k.a. Johannes Müller, whose preferred appellation was a self-consciously classicized version of the name of his home town, Königsberg. Regiomontanus was a mathematician and astronomer, but he was also a humanist specialist in Latin literature (especially Vergil). The attendant attitudes towards antiquity determined his other scholarly work, and appear with stark clarity in the *Epitome of the Almagest*, the work that he wrote with his older contemporary Georg Peurbach, another humanist and astronomer in Vienna.[2] The preface to this work, written in the early 1460s, is a humanist paean to the glories of antiquity and the contrasting cultural poverty of the present. Regiomontanus harangues his audience on the sad state of mathematical studies, and on the only proper way forward – the one that he promoted. He took this material on the road in the 1460s, most notably addressing university audiences at Padua in 1464 in a surviving lecture on the history of mathematics, the *Oratio introductoria in omnes scientias mathematicas* ("Introductory oration on all the mathematical sciences"). Regiomontanus's approach in the *Oratio* is designed to place him in a mathematical tradition that can be traced from the Egyptian origins of geometry, through the ancient Greek mathematicians and the successive translation of their work into Arabic and then Latin, to yield the present-day mathematical enterprise. This last, continuous with antiquity, is that in which Regiomontanus himself participated (again through forms of translation). And he regarded astronomy as the highest of the mathematical sciences.

Regiomontanus thus transferred the language of the humanists, on decline and renewal, to the specific arena of the mathematical sciences. The

Oratio and his preface to the *Epitome* are typical statements of humanist ideology, and can be found echoed in major texts of the sixteenth century that similarly address scientific matters. The *Epitome* had a large influence on astronomical training at the end of the fifteenth century, after its eventual printing in 1496, and its humanist rhetoric of "restoration" clearly found a receptive audience. Perhaps the most important of all the astronomical practitioners who followed in Regiomontanus's footsteps was the Polish canon familiarly known in Latin as Nicolaus Copernicus. In the early 1490s Copernicus studied at the University of Krakow, one of the new Polish universities of the fifteenth century and an institution with a quite vigorous tradition in astronomy. In addition to its astronomical status, however, Krakow had also become, by the end of the century, something of a centre for the new humanist learning, stressing the importance of the classical languages and the value of erudition in the texts of antiquity. Copernicus was studying in Italy when the *Epitome* came off the presses in Venice (he returned to Poland with some expertise as a medical practitioner, although he never took a medical doctorate). A few years later, in 1509, his first published work appeared: a Latin translation of a Greek poem. By around 1512, he had produced the first version of a new astronomical system intended to replace the world-system of Ptolemy. This text, known as the *Commentariolus* ("little commentary"), began with a consideration of Ptolemaïc astronomy and its supposed shortcomings; all the evidence suggests, however, that Copernicus's detailed knowledge of Ptolemy's astronomy at this time was still dependent on the account given in the *Epitome of the Almagest* rather than on Ptolemy's *Almagest* itself – a text that remained unprinted until a medieval Latin version from the Arabic was published in 1515, the original Greek text itself not appearing in print until 1538. The *Commentariolus* came to be known quite widely in astronomical circles during the sixteenth century, but only in manuscript form. The printed account of Copernicus's new system did not appear until much later, in 1543.

That later work, *De revolutionibus orbium coelestium* ("On the Revolutions of the Celestial Spheres"), which turned the earth into a planet that orbits the sun, was presented explicitly as a renovation of the ancient Greek astronomical tradition. In the preface to the work (dedicated to Pope Paul III), Copernicus tells of the route he had taken in arriving at his new ideas. They centred above all, as had his much briefer remarks at the beginning of the *Commentariolus*, on the shortcomings to be found in the current state of astronomical practice. Declaring that his intention was to improve matters, Copernicus makes the typical humanist move – he canvasses the available ancient authorities.

> I undertook the task of rereading the works of all the philosophers which I could obtain to learn whether anyone had ever proposed other motions of the universe's spheres than those expounded by the teachers of astron-

omy [lit. "mathematics"] in the schools. And in fact first I found in Cicero that Hicetas supposed the earth to move. Later I also discovered in Plutarch that certain others were of this opinion. I have decided to set his words down here, so that they may be available to everybody . . . [there follows a quotation from the Roman Plutarch's *Opinions of the Philosophers*].[3]

Copernicus's conclusion to this apparent digression is instructive. He says: "Therefore, having obtained the opportunity from these sources, I too began to consider the mobility of the earth. And even though the idea seemed absurd, nevertheless I knew that others before me had been granted the freedom to imagine any circles whatever for the purpose of explaining the heavenly phenomena."[4] Finding ancient precedent for the suggestion that the earth might move was essential to justifying his own consideration of the matter: it provided him with an "opportunity" that he would otherwise have lacked. Presenting one's work as innovative was seldom regarded as the best way to be taken seriously; innovations were light and insubstantial.

It needs to be stressed, however, that we cannot regard Copernicus's way of speaking as mere packaging. There is no basis whatever for thinking that Copernicus did not see his "new" astronomical system as a legitimate continuation of the ancient legacy represented by Ptolemy, and his own work as that of restoration. Furthermore, we have good evidence from a reliable source that this was indeed Copernicus's view of his own endeavour. The first printed discussion of Copernicus's sun-centred astronomy appeared in 1540, written by Georgius Rheticus, a mathematics professor from the University of Wittenberg. Rheticus (a Lutheran) had travelled to Thorn, or Toruń, in western Poland, to visit Copernicus (a canon in the Catholic church) in 1539. Rheticus was evidently drawn there by the high reputation that Copernicus had acquired over the years as a mathematical astronomer (he was never much of an observer) and by rumours of Copernicus's new astronomical system. Rheticus wanted to learn more, and the work of 1540, called *Narratio prima* (the "first account" of Copernicus's system), contains an outline of Copernicus's ideas and praise for their virtues – which for Rheticus amounted, when all was said and done, to their being *true*. Rheticus refers frequently in *Narratio prima* to the as-yet-unpublished text of *De revolutionibus*, which he was instrumental in persuading Copernicus to have published; in a letter he mentioned that Copernicus's great work had been written "in imitation of Ptolemy".[5] The word "imitation," unquestionably used by Rheticus as a term of approval, shows once again how Copernicus and his astronomical contemporaries viewed his work. Copernicus imitated Ptolemy just as a budding humanist rhetorician imitated Cicero; it was the way to acquire their skills. Hence Copernicus's greatest achievements in the eyes of his mathematical contemporaries lay in his consummate skill in devising geometrical models of

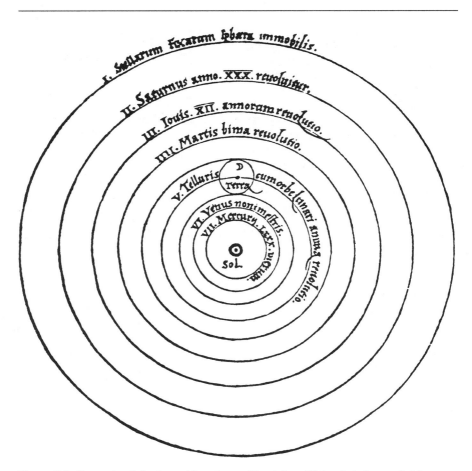

Figure 2.1 *Copernicus's basic world-system, without the additional circles needed for accuracy. From Copernicus,* De revolutionibus.

celestial motions using the same techniques as Greek astronomers themselves – including the firm restriction to uniform circular motions as the models' components. (See Figure 2.1.)

All this was despite the fact that Copernicus departed from Ptolemy radically, by setting the earth in orbit around a now-stationary central sun. This reformulation was more than simply astronomical, a new way to calculate the motions of lights in the sky; it also possessed *physical*, cosmological significance if taken literally. A moving earth that was no longer at the centre of the universe undermined many of the central tenets of

Aristotelian physics. As we shall see shortly, however, one way for the astronomer to stop short of drawing unwelcome physical implications from Copernican astronomy was precisely to *be* an astronomer, and *not* a natural philosopher.[6]

In pursuing a technical enterprise by hailing a return to the practices of the ancients, Copernicus was participating in the great scholarly cultural movement of his time, that of renaissance humanism. We should not, therefore, expect him to be alone in this kind of enterprise. The norms and conventions of humanist discourse conditioned not only the forms of presentation used in various of the sciences, but also the nature of those enterprises themselves, as we have just seen in the case of Copernicus. The goal of restoring modern society by returning to the cultural practices of antiquity could not be cleanly separated from the procedures of those sciences themselves. The "Scientific Renaissance," as we have called it, spans the sixteenth century precisely to the extent, and in the same way, that humanist education infused the scholarly perspectives and judgements of practically all those educated to anything near university standards.

III Finding out how the ancients did it

The anatomist and physician Andreas Vesalius of Brussels affords another striking instance of a renowned figure in the history of science who must be seen as a part of this same cultural movement. Vesalius is best known for his influential publications – printing being, again, an integral part of the story. His greatest work is *De humani corporis fabrica* ("On the Fabric of the Human Body"), published, like Copernicus's *De revolutionibus*, in 1543. Vesalius had been trained as a physician at the University of Paris, and subsequently taught at the universities in Louvain, Paris, and then Padua. Most of the specifics of what can be said about his early career depend on his own account of them, an account that was itself fashioned with particular, interested aims in mind. Fortunately, enough is known independently about the universities with which he was associated to allow checks on some of his claims. But one of the things of which we can be sure is that Vesalius had the technical skills and the intellectual sympathies required to make him a humanist scholar.

Vesalius was born in 1514. In the 1530s he was drawn into collaborating on the production of a new, scholarly edition of the works of the pre-eminent medical authority of antiquity, Galen, an edition that appeared in 1541. This edition was intended to publish Galen's many Greek texts in good new Latin versions, with all the appropriate scholarly apparatus commenting on and explicating the philological niceties of Galen's language and terminology. Such an enterprise stood at the centre of humanist scholarship; classical culture could not be revived without intimate understanding of the sources. In introducing the *De fabrica*, Vesalius availed himself of the same humanist sensibilities. That book is presented, in a dedicatory preface to the Holy Roman Emperor, Charles V, as a contribution to the general restoration of learning. Much like Copernicus on astronomy, Vesalius starts out by bemoaning the decadent state of contemporary medicine and of its decline since antiquity. He then speaks of how, in the present age, when "anatomy has begun to raise its head from profound gloom, so that it may be said without contradiction that it seems almost to have recovered its ancient brilliance in some universities," he had thought it time to write this book to assist that process of revival.[7] Vesalius explains his goal like this:

> I decided that this branch of natural philosophy ought to be recalled from the region of the dead. If it does not attain a fuller development among us than ever before or elsewhere among the early professors of dissection, at least it may reach such a point that one can assert without shame that the present science of anatomy is comparable to that of the ancients, and that in our age nothing has been so degraded and then wholly restored as anatomy.[8]

The touchstone of Vesalius's reverence for the medicine of antiquity was Galen. Galen had been the pre-eminent medical authority of the Middle Ages, just as Ptolemy had been the chief astronomical authority; but just as the humanist revival in astronomy focused on restoring Ptolemy's entire enterprise instead of simply making pragmatic use of his results, so with the humanist medical project the object was to restore the *kind of medicine* that Galen had written about, not simply to parrot his words as if they were ultimate authorities on all points of fact and interpretation. Thus, in the passage quoted above, Vesalius speaks of reviving a method of dissection that could stand comparison with that of the ancients. Vesalius, as an anatomist, was particularly concerned with stressing Galen's own works on that subject as well as his associated expertise as a surgeon. It is noteworthy that Vesalius has here spoken of anatomy as a "branch of natural philosophy," because that characterization emphasizes the contemplative aspect of the subject rather than its practical significance, which Vesalius had pointed out elsewhere in the preface. And indeed, Galen's own approach to anatomy had been deeply informed by Aristotelian philosophical conceptions, in themselves aimed at understanding rather than at operation.[9] While acknowledging the errors to be found in Galen's anatomical writings, Vesalius still upholds him as the model to be emulated, even to the extent of organizing the *De fabrica* according rather to the principles of Galenic philosophical anatomy than to the conventional practice of his own time. Galen had argued that the topical treatment of the fabric of the human body should proceed from the outer parts (veins, arteries, muscles and nerves) and only then move inwards to the viscera. Vesalius followed Galen, starting out by explicating the skeletal structure of the body as its basic framework.[10] Demonstration-dissections in Vesalius's time, by contrast, routinely displayed the viscera first of all, simply because they decay more quickly than other parts; this cuts no ice with Vesalius, because he is interested in presenting anatomy as a "branch of natural philosophy," and has "followed the opinion of Galen,"[11]

Unlike Copernicus, Vesalius did not propose major changes in the theoretical framework of his science in the name of restoration. He did note that the moderns were better able to investigate human anatomy than Galen had been, because Galen had relied on inferences from dissections of apes, not of human cadavers. On that basis, Vesalius was happy to show that Galen could be corrected on specific points. But he was not concerned with undermining the broader medical system that Galen represented, to do with the causes of disease and with the physiological workings of the body. Anatomy was basically a descriptive science, speaking of the body's construction; it was not centrally concerned with the operations of the body, causally understood. Copernicus worked within an astronomical tradition that had traditionally stood somewhat apart from issues relating to the physics of the heavens, but his astronomical innovations carried with them serious challenges to Aristotelian (and Ptolemaïc) cosmology. Vesalius presented his own work as part of an anatomical tradition, but did not attempt to question Galenic physiology on its basis – despite describing it as a "branch of natural philosophy," perhaps to inflate its dignity. (See Figures 2.2 to 2.5.)

Vesalius's humanist approach to his specialty is not only unsurprising, in light of the cultural setting that had made him, but also consequential. Like Copernicus, Vesalius presented his work as restoration of an ancient practice; also like Copernicus, he pointed out flaws in the work of his great model from antiquity; and like Copernicus, the rationale for his project emerged directly from humanist values and ambitions. Several centuries of learned anatomical knowledge had not seen fit to attempt the revival of Galen's enterprise; now Vesalius promoted it, and through the medium of print he disseminated it. We should note a further feature of his work, however: above all, he is renowned as having advocated a return to hands-on anatomical research, rather than assuming that Galen was always right. While his own self-presentation in this light was certainly somewhat misleading (he was not, for instance, the first professor to perform personally the demonstration-dissections shown to students, as he made it appear in *De fabrica*), he was certainly a part of a newly-vital tradition of research in anatomy that credited its inspiration to Galen's example, and that continued after him at the University of Padua down into the seventeenth century.

It would be tedious to enumerate every case in which the values of humanism infused the practices of the sciences in the sixteenth century. Copernicus and Vesalius are worth noting, however, simply because they are such major figures in our understanding of the period; they cannot be

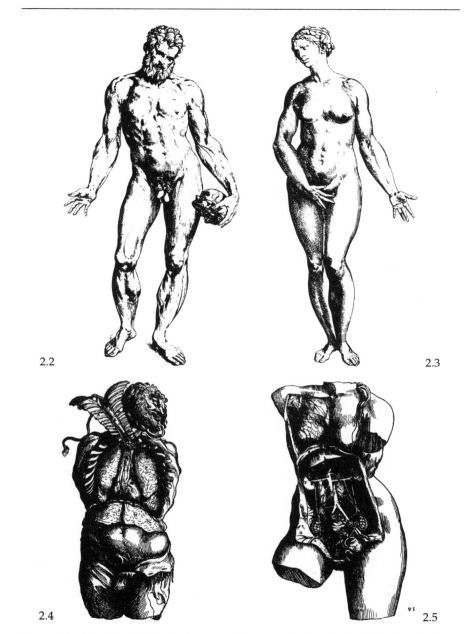

2.2

2.3

2.4

2.5

Figure 2.2, 2.3, 2.4, 2.5 *Idealized man and woman, in the classical Greek style (from* Vesalius's Epitome *of the* De humani corporis fabrica*), and surgically butchered male and female cadavers (from Vesalius's* De humani corporis fabrica*). The first two represent humanistic physical and cultural perfection, while the second two represent a fully materialized anatomy of suffering.*

dismissed as mere curiosities. One final example will perhaps serve to bring the point home. One of the architects of modern symbolic algebra was the French mathematician François Viète (1540–1603). His achievements towards the end of the sixteenth century were not put forward as the product of an original mathematical mind, however; his greatest work, in 1600, was called *Apollonius Gallus* ("French Apollonius"), to signal to all that he was emulating the achievements of the third-century BC Greek mathematician and astronomer Apollonius. Viète, like many other mathematicians of his time, was convinced that the ancient Greek mathematicians had known a form of "analysis" in geometry that had enabled them to discover the theorems for which they subsequently provided deductive proofs from first principles. It was a source of incredulity to many that the Greeks could have found so many counter-intuitive results if they had not possessed an "art of analysis" by which to do it – everyone knew that it was much easier to prove a result that you already knew. Some scant textual evidence fuelled this idea, most notably the late-antique text known as the *Arithmetic* (precise period of composition unknown), by one Diophantus of Alexandria. Diophantus's work contained techniques for finding unknowns which are, in hindsight, best understood as algebraic; but they appear in the form of worked examples using actual numbers, and amount to practical computational methods rather than part of a theoretically-grounded branch of mathematics. It was the attempt to develop into a properly-grounded branch of mathematics both Diophantus's approach and established mercantile calculatory techniques (known as "the art of the coss")[12] that yielded what Viète called the "art of analysis." He presented it as a reconstruction of the "art" whereby the ancient Greek mathematicians had routinely found their results. In the earliest decades of the following century, other mathematicians continued to adhere to this "reconstruction" approach to analysis. Even René Descartes, the inventor of the modern symbolic algebra with which we are today familiar, believed early in his career (1620s) that the ancients had indeed possessed such an art, but that "they begrudged revealing it to posterity."[13]

IV Renovation, innovation, and reception

The new reformed Lutheran universities of sixteenth-century Germany illustrate some of the flexibility that could attend the reception of such developments. Copernicus's follower Rheticus, at the time of their first personal encounter, was a professor of mathematics at the University of Wittenberg; shortly thereafter he took up a similar post at the University of Leipzig. Both were Lutheran institutions, Wittenberg the first among them, and both were important sites for the reception of Copernicus's *De revolutionibus* by academic astronomers. Wittenberg's group of astronomers was of especial significance in training new astronomers and in disseminating particular approaches to the science.

Rheticus, despite his role in bringing *De revolutionibus* to publication, was not around to see the book through the press (it was printed in Nuremberg). Instead, for reasons that are obscure, oversight was given to a Lutheran theologian, Andreas Osiander, who took the opportunity to add to the book's front-matter a short, unsigned preface which many subsequently took for Copernicus's own. The book also included Copernicus's dedicatory preface to the Pope. The Lutheran's addition did not amount to a theological conflict with Copernicus's views; rather, it was concerned to defuse a potential problem that stemmed from the way in which Copernicus had spoken of his system both in the dedicatory letter and in the body of the work itself. That problem related to the issue considered in the previous chapter: the physical status of astronomical models.

There is no doubt that Copernicus saw his astronomical system of the universe, with its moving earth and stationary sun, as a true representation of how the cosmos was actually constructed. Book I of *De revolutionibus* is concerned, in imitation of the *Almagest's* first book, with establishing the basic architecture of the universe. Where Ptolemy had used basically Aristotelian physical arguments in support of the doctrine of a central, stationary, spherical earth around which the heavens revolved, Copernicus had to make up a few physical principles of his own to make the basic structure of his own universe at least plausible. Just as Ptolemy had imported such things into astronomy from outside, so too did Copernicus. Hence questioning them would not in itself affect the properly mathematical, astronomical components of the system; it only bore on their interpretation as representations of reality. But despite the lower *astronomical* importance of these physical considerations, Copernicus clearly took them very seriously, because without them he would have had no basis on which to portray his system as *true*.

It is nowadays regarded by historians as well-established that Copernicus not only regarded his heliostatic[14] world-system as physically true in its general outlines, but also, in principle, in its operations. That is, Copernicus appears to have credited the movements of the celestial bodies to the revolution of physically real spheres in the heavens, which carried them around. In this sense, he was following directly in the footsteps of Peurbach in the *New Theorics of the Planets.*[15] Where Peurbach had represented the circles of his Ptolemaïc planetary models as if they were physically real bodies to which their visible passengers were attached, Copernicus seems, almost by default, to have assumed something very similar for the circles in his new non-Ptolemaïc system. After all, something needs to account for the movement and the paths of the celestial bodies through space – but only if the relevant astronomical models under consideration are regarded as physically true representations that *explain* celestial motions.

The astronomers at the University of Wittenberg balked at taking this step: they held fast to the disciplinary division separating mathematical astronomy from the *physics* of the heavens; the latter was no real concern of theirs. Thus, in adopting *De revolutionibus* in their teaching and practice of astronomy, the Wittenbergers cheerfully ignored its physical, cosmological pretensions. Following the standard Ptolemaïc model, however, they usually started out their textbook courses on astronomy with a brief discussion of the reasons for believing in a stationary, central earth, and the other, similar points that appeared in the *Almagest* (including Ptolemy's arguments against the motion of the earth). Thereafter, they were free to use Copernicus's geometrical models as they saw fit, in keeping with the spirit of Osiander's anonymous preface, entitled "To the Reader Concerning the Hypotheses of this Work."

Osiander had taken pains to stress that Copernicus's astronomical system, as detailed in *De revolutionibus*, should *not* be taken as a representation of physical reality. Instead, he says, the proper function of the astronomer is to collect observational data and devise hypotheses for them which will enable "the motions to be computed correctly from the principles of geometry for the future as well as for the past":

> For these hypotheses need not be true nor even probable. On the contrary, if they provide a calculus consistent with the observations, that alone is enough.... For this art, it is quite clear, is completely and absolutely ignorant of the causes of the apparent nonuniform motions. And if any causes are devised by the imagination, as indeed very many are, they are not put forward to convince anyone that they are true, but merely to provide a reliable basis for computation.[16]

Astronomy, says Osiander, is completely ignorant of the laws *producing* apparently irregular motions; in other words, its proper laws are laws relating to the *description* of celestial motions rather than laws proving causal *explanations*. This is a stark statement of the disciplinary divide between astronomy and cosmology noted in the previous chapter; Osiander treats it as absolute, in contrast, for example, to Peurbach – or Copernicus himself – who had tended to blur the boundary. On this basis, Osiander advises the reader not to take seriously Copernicus's foundational hypotheses of the earth's motion, lest he "depart from this study a greater fool than when he entered it."[17]

Osiander's intention, it is usually assumed, was to protect Copernicus from criticism on theological grounds (scriptural passages could be, and very swiftly were, presented that appeared to indicate the earth's stability). That would be a piquant situation, to have a Lutheran theologian attempting to shield a Catholic canon. But in fact, as the foremost historian of Copernicanism, Robert Westman, has shown, Osiander's remarks are fully consonant with the practice of the Lutheran Wittenberg astronomers. Men such as Erasmus Reinhold and Caspar Peucer in the middle decades of the century praised Copernicus's astronomical achievements while discounting his central cosmological thesis. Copernicus's mathematics could be used for predicting motions as they appeared in the sky without assuming that the earth was actually moving. It was just a matter of recognizing that one can switch from a reference-frame in which the earth moves but the sun is fixed to one in which the sun moves and the earth is fixed; all the relative motions remain the same.[18]

The wider importance of Wittenberg in this story is connected to the fact that Peucer, in particular, trained students who themselves subsequently went out to spread these approaches to astronomy at other German universities. The University of Wittenberg was the flagship of the newly-Lutheran universities of mid-century, acting as something of a model for others. The person most responsible for the reform of the curriculum at Wittenberg, to make it consonant with Lutheran views, was Martin Luther's scholarly right-hand man Philip Melanchthon. The name enables us once again to identify humanism as a centrally relevant cultural dimension of Melanchthon's activities. It comes from the Greek, meaning "black earth," and translates the original German name *Schwartzerd*. This is the same phenomenon that we have already seen in the case of Johannes Müller, "Regiomontanus," with the difference that Melanchthon uses Greek to designate his own allegiance to the ideals of classical culture where Regiomontanus had been content with Latin – knowledge of Greek was by now a lot more widespread among scholars than it had been just a few decades earlier.

Accordingly, Melanchthon's curricular reforms at Wittenberg from the 1520s onwards emphasized classical learning in the humanist style at the expense of some features of the old scholastic learning. In particular, Melanchthon pressed for natural philosophy to be taught not from Aristotle but from the elder Pliny's *Natural History*. The latter, a Roman text from the first century AD, recommended itself to Melanchthon on a couple of different grounds. One was simply that it was not Aristotelian. This is not to say that Melanchthon had no regard for Aristotle: Aristotle was, after all, a learned ancient, whose philosophical and logical writings had to be taken seriously by any scholar. But Melanchthon wanted to banish the *scholastic* Aristotle and his dominance of the old university curriculum; throwing out use of many of the texts themselves was the most radical way to do that. The other important reason for preferring a natural philosophy based on Pliny rather than Aristotle was that Pliny discusses matters of *practical*, operational significance. Where Aristotle is concerned to provide theoretical understanding mostly divorced from practical applications, Pliny by contrast gives techniques for making dyes or mining ores. A shift towards Pliny therefore implied a shift towards a different, operational conception of natural philosophy.

In the event, Aristotle's natural philosophical writings proved to be indispensable for any serious academic programme of study in areas as diverse as physics and psychology (the latter following the long tradition of commentaries on Aristotle's *On the Soul*). Too much of academic scholarly life had been bound up with Aristotelian texts for too long to cast them aside; they contained too much of importance. But, in keeping with Melanchthon's humanist predilections, Aristotle began to be studied with much closer concern for fidelity to the original Greek text and to related philological matters. The original Aristotle was just as important to recover as the original Ptolemy – or the uncorrupted text of the Bible.

V Restoration and a new philosophical programme: *Archimedes redivivus*

There is scarcely a single branch of learning in the Renaissance that was unaffected by the humanist cultural movement. Much as mathematicians like Viète strove to recover the analytical capabilities of antiquity, botany too was revived through intensive work at identifying with actual species the plants described in ancient works by such as Aristotle's pupil Theophrastus. The route to cultural respectability tended to lie along the path of identification with ancient authority, and it constituted a recipe that many people found it advantageous to follow. While we have so far considered only those who appealed to ancient texts whose authenticity remains unquestioned to this day, there were also people who appealed to the authority of writings that were subsequently shown to be less than had been claimed. The previous chapter considered briefly the hermetic corpus, the body of late antique texts that Ficino translated from Greek into Latin. The hermetic writings were claimed to date from remote antiquity, but in the early seventeenth century the classical scholar Isaac Casaubon established them as much more recent. The credibility attaching to many forms of magical practice owed a great deal to the authority of these texts, but that authority faded only gradually after Casaubon published his results.

For a small but significant group of scholars in the sixteenth century, the works of the Greek mathematician Archimedes were another such source of inspiration, legitimation, and example. If one could portray one's own work as implicit in, continuous with, or having precedent in the work of an ancient (as we saw Copernicus attempt to do for the doctrine of the earth's motion), that new work would immediately appear more respectable and hence more *likely*. Once again, it should be stressed that there is no reason to suppose that these associations with ancient authorities were cynically conjured up to sell new ideas to an unsuspecting world; no doubt Copernicus really was pleased that he could find ancient precedent for a moving earth, just as Vesalius was genuinely taken by what Galen had to say about the performance of one's own dissections. But the upshot of this situation was that the characteristic form, in practice, through which new directions were pursued in learned activity at this time was that of identification with the precepts and example of a thinker from antiquity – preferably one who wrote in Greek or good Latin.

Printing continued to function as an important medium through which such programmes could go forward. In the 1550s Latin translations of some of the works of Archimedes (third century BC), known in Latin in the Middle Ages but little used, were printed in Italy in revised, restored versions under the editorship of Federico Commandino. They were swiftly adopted as appropriately classical models in the practical arena of mechanics: Archimedes' two works known as *On the Equilibrium of Planes* and *On Floating Bodies* presented formalized mathematical sciences that related directly to mechanical devices for making work easier – machines in the classical sense. *On the Equilibrium of Planes* concerns the behaviour of levers and balances, as a prolegomenon to theorems concerning the determination of the centres of gravity of various kinds of plane figures. *On Floating Bodies* examines the conditions under which solid bodies will float or sink in liquid media, in regard to the specific gravities of each. Both of these texts offered means for demonstrating with mathematical precision instrumental techniques relating to mechanical situations, but their significance was augmented by the image of Archimedes himself as presented some centuries after his time by the Roman Plutarch. Plutarch told of an Archimedes who assisted the king of Syracuse (the Greek colony city in Sicily in which Archimedes lived) in the capacity, in effect, of an engineer, aiding in the city's defence against the Romans. Archimedes could therefore stand as an ancient exemplar of the learned, yet also practical, engineer, and the adoption of Archimedean kinds of demonstrations in mechanics became the hallmark of a group of northern Italians in the second half of the sixteenth century.

Commandino, who had embarked on a programme of printing all of Archimedes' surviving mathematical works in Latin versions so as to bring out the virtues and illuminate the obscurities hidden in the medieval Latin translations, began this Archimedean revival in the Italian city of Urbino. He was followed by two figures of particular note, Guidobaldo dal Monte, a nobleman; and Bernardino Baldi. Baldi wrote a history of mathematics that, among other things, reconstructed the history of mechanics using Archimedes as its pivotal point, thereby attempting to establish a tradition into which Italian mechanicians like him could fit. Guidobaldo is best known, besides his intellectual accomplishments, for having been an early and influential patron of Galileo. It was under his influence that Galileo himself adopted the tenets of these "philosopher-engineers," which appear very strongly in his earliest writings from around 1590. Galileo there concentrates on undermining Aristotelian physical authority on matters of moving bodies, in favour of an Archimedean-style mathematical approach. In other words, while rejecting the contemporary (and corrupted) authority of Aristotle, this school of practitioners trumpeted the authentic antique authority of Archimedes. At the same time, they effectively challenged the assumption that intellectual knowledge was categorically distinct from practical capabilities.

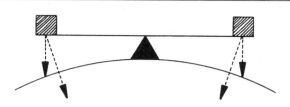

Figure 2.6 *The strict lack of parallelism between the potential (unconstrained) lines of descent of the weights at opposite ends of a balance.*

Familiar features appear also in this case: using an ancient authority as one's model did not mean a slavish adherence to everything that authority had said or done. Guidobaldo dal Monte was just as ready to criticize aspects of Archimedes' work as Copernicus had been of Ptolemy, or Vesalius of Galen.[19] Here, the criticism struck at the core of Archimedes' work on balances. In *On the Equilibrium of Planes*, Archimedes had conceived of the weights on opposite ends of a simple balance as tending downwards in straight lines that were parallel to one another (see Figure 2.6). Guidobaldo objected that in fact (as Archimedes himself would have known), heavy bodies exert themselves towards the centre of the spherical earth (or universe). Thus the ends of the balance should tend downwards along lines that, instead of being parallel, would, if prolonged, *intersect* at the earth's centre. Archimedes' demonstration of the law of the lever (i.e. that a balance is in equilibrium when the weights on its two arms are in inverse proportion to their respective distances from the fulcrum, or pivot-point) used this false assumption of parallelism throughout; Guidobaldo therefore condemned it as unrigorous and wished for an improvement.

Notice that this was not a matter of practical significance. Any real situation with a normal balance would involve a deviation from parallelism far too small to be measurable; and yet Guidobaldo still worried about it. The ancient Greek mathematical enterprise, of which Euclid was the prime example and Archimedes one of the greatest exponents, placed great stress (as had Aristotle more generally) on the virtues of rigorous, absolute demonstration, and its apparent absence in this case evidently bothered Guidobaldo more than it had Archimedes himself. The Italian philosopher-engineers were attracted by the lure of a formal scientific mechanics that could hold up Archimedes as its hero, but theirs was not intended to be an antiquarian enterprise of celebration; it was, once again, an effort at emulation. Guidobaldo, said Baldi, was responsible for the "restoration of mechanics to its ancient splendour."[20]

Together with Viète's quest for the ancient art of analysis, the

Commandino–Guidobaldo–Baldi Archimedean revival is perhaps the last important example of the "Scientific Renaissance". Strong convictions about the superior wisdom of antiquity began to give way in the new century to a subtly altered perspective. Viète's ambition had been to "equal and surpass" the ancients by learning their own game and playing it better than they had, just as Copernicus wanted to improve on Ptolemy by playing what he considered to be the same game better, and Vesalius wanted to improve on Galen by carrying on the enterprise of anatomy under improved conditions. But in the early seventeenth century, people increasingly wrote of making a clean break with the past, as we shall see most starkly in the cases of Francis Bacon and René Descartes. Descartes, as noted above, had in the 1620s believed in the existence of an ancient mathematical art of analysis that had been lost. It was only after reviewing the new competences of his own analytic art in his famous essay of 1637, the "Geometry," that he declared that he had invented something *new*, something that the ancients had surely not possessed. By the end of the seventeenth century, the so-called "battle of the books" in England as well as elsewhere typified a new situation: while controversy could still rage over the comparative literary merits of ancient and modern poetry, scarcely anyone was prepared by then to deny that recent developments in the sciences had leaped far ahead of the ancient legacy.

Why this change in perspective had occurred is unclear; it may have been a result of the perceived achievements that had sprung from the restoration efforts of the sixteenth century, or it may have come about as more scholars came to realize that the legacy of classical antiquity was diverse. Different ancient authorities said contradictory things about the same topics, as the welter of readily available printed editions easily made manifest. It became harder and harder to identify an ancient orthodoxy to be restored. Ancient texts continued to be enormously important resources, but no longer as signposts to a past golden age.

Chapter Three
The Scholar and the Craftsman: Paracelsus, Gilbert, Bacon

I Mastering the occult

The restoration of ancient culture was just one of the preoccupations of sixteenth-century discussions on the knowledge of nature. Outside those arenas in which the university-educated paraded their humanist credentials, voices began to be raised against the dominance of scholastic values in learning. In particular, the usual Aristotelian emphasis on contemplative rather than practical knowledge of nature came in for severe criticism, usually on moral grounds. In Greek, this distinction was denoted by the terms *epistēmē* and *technē*, corresponding to the Latin *scientia* and *ars* ("science" and "art"). The school stress on *scientia* appeared to some critics as a deliberate neglect of practical matters, being of especial culpability in the case of medicine, in which practical ends were most obviously at issue.

The starkest example of these views in the first half of the sixteenth century appears in the work of Paracelsus. The German medical mystic Philippus Aureolus Theophrastus Bombastus von Hohenheim traded in his impressive name for the punchier "Paracelsus" probably as a means of advertising his claim to have gone beyond the abilities of ancient physicians, represented here by the first-century Roman Celsus. Paracelsus spent his life travelling the German territories of central Europe (especially Switzerland) promulgating his cosmological doctrines and their medically efficacious implications.

It was a central part of Paracelsus's teachings that a true knowledge of the natural world, on which medical treatment should be based, could only be acquired through an intimate acquaintance with the properties of things. In this, he reflects a longstanding, if somewhat unorthodox, position that had been represented in the thirteenth century by the English friar Roger Bacon, who had advocated something called *scientia experimentalis*. This was a romantic notion aimed at creating a certain oneness between the would-be knower and the object to be known, and in Paracelsus's case it was explicitly spiritual and alchemical. Paracelsus's importance in the history of *materia medica* stems precisely from the fact that he and his followers advocated use of various inorganic chemical (i.e. mineral) substances to treat diseases. Paracelsus argued for the use of these new medicines, while decrying the inefficacy of the usual Galenic doctrines taught in the universities, not by claiming their greater authenticity as representatives of a pristine classical medicine, as Vesalius might have done, but by claiming an essential *novelty* for them. These times, he said, have brought forth new and virulent diseases. The spread of syphilis in sixteenth-century Europe (once suspected to have been brought back from the newly opened-up Americas) was perhaps the most evident example. In the face of new diseases, new remedies were therefore called for. In effect, Paracelsus announced a break with the past: a new world of disease and a new world of medicine to confront it.

The medical faculties of the universities had, since their foundation in the thirteenth century, taught according to the teachings above all of Galen and the Arabic philosophers Avicenna and Rhazes.[1] The two latter had written works on medicine that observed the main lines of Galen's theoretical doctrines, so that the overall approach in the western Middle Ages is appropriately labelled "Galenic." The central therapeutic doctrine concerned the balance of the four "humours" that comprised the human body: blood, phlegm, yellow bile and black bile (this last not identifiable with any modern physiological entity). A preponderance of any one of these humours corresponded to a characteristic disposition: an individual with a preponderance of blood was sanguine, of phlegm phlegmatic, of black bile melancholic, and of yellow bile choleric. Too much of an imbalance of humours brought about a pathological condition, an illness, that the physician attempted to cure. The basic means of doing this was to counter the particular combination of qualities associated with each humour. One of the properties of blood, for example, was that it was hot; a fever, regarded (owing to its heat) as being due to an excessively sanguine condition, was therefore best treated by attempting to cool the patient.[2]

Paracelsus poured scorn on such ideas. In their place, he spoke of sympathies between different parts of nature, such as alchemical correlations between particular planets and particular minerals. These correlations extended, crucially, to the human body itself. Knowing the appropriate occult sympathies between the parts of the body and the "virtues" of things in the world (such as particular herbs or metals) enabled a determination of the correct treatment of some ailment. In setting up the justifications for such procedures, Paracelsus thus adopted the classical microcosm/macrocosm analogy inherited from antiquity. On this view, the human body is a mirror of the universe as a whole, a "microcosm" or small world reflecting the great world, the "macrocosm." Each part of the heavens in Paracelsus's geocentric universe (the five planets, sun, and moon) was deemed to have its correlate in the human body. Paracelsus's language here spoke of *astra*. An *astrum* was a virtue with its prototypical representation in the heavens (associated with a particular planet, for example) but with its correlate in the human body. Thus "a wound below the belt contracted when the moon is new is unluckier than one contracted when the moon is full," while "wounds contracted under Gemini, Virgo, Capricorn are the most unlucky."[3] Paracelsus's use of astrological ideas and categories went along, however, with an explicit repudiation of astrology itself: where astrology as Paracelsus understood it spoke of the causal *influence* of the heavens on terrestrial affairs, he himself, in what he called his "astrosophy," saw only correlations or correspondences between these parts of nature. *Astra* were also to be found with non-human terrestrial things (plants and minerals, typically), thereby providing a basis for medical treatment. The pre-existing alchemical associations between particular celestial bodies and particular metals, such as copper and Venus or iron and Mars, facilitated the identification of such correspondences.

But just as Paracelsus borrowed from astrology while making consider-able modifications to it, so also he modified the traditional teachings of alchemy. Alchemy had entered the thought of the Latin Middle Ages from Arabic sources. Alchemists thought in terms of the four elements accepted in ancient Greek natural philosophies, including that of Aristotle: earth, air, fire and water. Paracelsus, although very critical of contemporary scholas-tic Aristotelianism, did not reject these elements; but he did supplement them in a rather ambiguous fashion. His favoured alchemical fundamen-tals were the *tria prima*, namely salt, sulphur, and mercury, which he des-ignated "principles" rather than "elements." Their chief function seems to have been as property-bearing constituents of bodies; ways, therefore, of designating the basic characteristics of a particular substance. An inflam-mable body, for example, might for that reason be regarded as sulphurous: the fire that is revealed by the process of burning shows the body's true composition. The four Aristotelian elements were, for Paracelsus, not truly elementary but, rather, representative of the material husks that contained a body's true and active spiritual essence.[4]

The doctrines of Paracelsus and his many followers, while often obscure, are clear in certain basic respects. Paracelsus rejected official school doc-trines (except insofar as, unavoidably and unintentionally, he adopted some of them himself); he emphasized the direct interrogation of nature as the route to knowledge; and he regarded knowledge of nature as pre-eminently practical and operational. Death from common medical ailments was a constant of life in early-modern Europe, and medicine was always one of the most prominent examples of concern with utilitarian knowledge. For Paracelsus, the new seekers after truth in nature were not to be scholars, but ordinary people in touch with the natural world. Thus Paracelsus's writings appeared (most of them after his death in 1541) in German rather than Latin, the shared learned language of Western and

Central Europe, although Latin translations were soon available. Paracelsus's later followers, such as Oswald Croll in the late sixteenth century and Johannes Baptista van Helmont in the seventeenth, tended to be clustered disproportionately in the Germanic territories of Europe but could be found all across the continent. Paracelsus's legacy was thus adopted by those who were attracted by his message of practical knowledge of nature in the service of medicine, underpinned by a magical/alchemical view of nature that saw the physician as a kind of magician who manipulated the sympathies and correspondences that knitted the world together.[5]

II Craft knowledge and its spokesmen

Paracelsianism was but one manifestation in the sixteenth century of a growing sense that nature, to be understood, needed also to be mastered – that those who truly *knew* nature necessarily also *commanded* it. Those representatives of literate culture who began to make such arguments were often placed in a strange social position. Like Paracelsus, they were advocating a much higher cultural prestige for the practical know-how of artisans and craftsmen, people usually far below them on the social scale. In order to make their claims effective, therefore, they needed to attempt to raise that prestige. In Chapter 2 we saw that the "philosopher-engineers" of the Italian renaissance glorified the work of engineering and mechanics by associating it with the name of Archimedes. Other spokesmen engaged the problem more directly, stressing the importance of this kind of active knowledge rather than the relative unimportance of most of its practitioners. In doing so, they could set themselves up as prophets of a kind of value-added practical knowledge wherein the untutored artisan would be disciplined by the literate gentleman overseer.

There are several prominent examples of such work in the sixteenth century. Vanoccio Biringuccio's *Pirotechnia* (1540) was a work on the business of mining and metallurgy written in the vernacular; in 1556 (a year after its author's death) there appeared another treatise on the same subject called *De re metallica* ("On Metals") by Georgius Agricola (Georg Bauer), a German mining engineer in Saxony. Agricola's work laid out in detail, this time in an elaborate Latin text, knowledge associated with practical matters of the mining of metallic ores and their refining. Agricola wrote according to humanist models in a book clearly aimed at the educated élite, attempting in effect to make mining a proper pursuit for a gentleman by associating it with such high-class ancient repositories of knowledge as Pliny's *Natural History*, and equipping it with a Latin technical vocabulary.[6] In the second half of the century, assertions of the importance and value of such practical matters had become a commonplace, whether associated with vernacular texts aimed at the less learned and tending to expose traditional craft secrets, or with Latin texts directed towards the educated élite and aimed at increasing the social status of such knowledge.

Indeed, in the latter case, there was a convenient classical category that could lend additional respectability to practical kinds of knowledge about nature. Ancient as well as Christian moralists had written of the distinct virtues and disadvantages of the *vita contemplativa*, or contemplative life, and the *vita activa*, or active life. The first was a life devoted to the improvement of one's own soul through solitary meditation and reflection, while the second stressed social engagement and involvement in civic affairs. Such categories were easily adapted to considerations of the use and purpose of knowledge about nature, of whether it should be about understanding nature or about the practical exploitation of nature's capacities. Andreas Libavius, an important writer on chemistry at the beginning of the seventeenth century, stressed in his work the civic role of the chemist and the importance of being an active participant in the affairs of one's community. He drew stark, explicit contrasts with the secretive labours of the closeted alchemist. One of the many notable features of Libavius's great textbook of chemistry, the *Alchemia* (1597), is its lengthy presentation of the types of apparatus employed by chemists, reminiscent of Ptolemy's discussions of astronomical instruments in his *Almagest*. Clinking glassware was becoming respectable.

The art of navigation was of especial concern in these centuries of the expansion of European trade around the globe, and by the end of the sixteenth century practical navigators had developed much knowledge of how to manoeuvre ships over enormous distances without getting too lost. Their practical mathematical techniques, as also those of landbound surveyors, served as models of useful knowledge that appealed to many promoters of the value of craft know-how. A sign of these changing attitudes was the inauguration in 1597 of Gresham College in London, an institution intended to provide instruction to sailors and merchants in useful arts, especially practical mathematical techniques. Late-sixteenth-century England was, indeed, rife with such projects, often aimed at strengthening the state by improving the abilities, especially commercial, of its people. These ambitions were matched by the appearance of many printed books containing practical mathematical techniques used in navigation and land-surveying.[7]

The relationship between the work of such mathematical practitioners and the speculations of philosophers about the physical world was occasional rather than intimate, but it did exist. Thomas Digges was the son of Leonard Digges, a mathematical practitioner who was the author of several books of this kind; Thomas himself was concerned with the same matters. In his 1576 re-edition of an almanac produced by his father, Thomas included an appendix of his own entitled *A Perfit Description of the Caelestiall Orbes*. This amounts to a cosmological discussion of the new

Figure 3.1 *Water-powered heavy industry in the sixteenth century, from Agricola's* De re metallica.

Copernican system of the universe, loosely based on Book I of *De revolu-tionibus*. Its qualitative character took off from a nuts-and-bolts treatment of practical (calendrical) astronomy, Digges apparently seeing no conflict in the juxtaposition. He shows clearly the potential of mathematics to speak to natural philosophy.

A further example of the mixing of mathematical practice and natural philosophy appears in the work of another Englishman, William Gilbert, who published in 1600 a celebrated work *De magnete* ("On the Magnet"). Gilbert was a physician (at one time in the service of Queen Elizabeth herself), and his book is notable upon several counts. First of all, in a gesture increasingly to be found among advocates of forms of empiricism, Gilbert was scornful of established Aristotelian learning. Ostentatiously rejecting tradition, he proposed to learn truths about nature through first-hand examination of things themselves. Secondly, much of the natural philosoph-ical content of *De magnete* owes considerable debts to sixteenth-century writers in broadly magical or animist traditions, especially Girolamo Cardano. Gilbert viewed the earth not just as a giant magnet (a contribution in itself), but also as in some sense alive and self-moving, at least around its own axis if not around the sun – he is cagey about the latter point. The Stoic-influenced philosophy of Cardano was a major resource for Gilbert in buttressing this view. Thirdly, Gilbert's approach to his arguments con-cerning the properties and behaviours of magnets was both deliberately experimental, involving careful tests of alleged properties of magnets, and informed more generally by the lore of seamen. The use of the magnetic compass for navigation had become well-established by the sixteenth century, and the accumulated experience of sailors in its use provided Gilbert with a stock of concepts, instruments, and alleged magnetic pro-perties to use and to test in his experimental investigations.

His debt to such know-how, furthermore, is an explicit one. He acknowl-edges those

> who have invented and published magnetic instruments and ready methods of observing, necessary for mariners and those who make long voyages: as William Borough in his little work the *Variation of the Compass*, William Barlowe in his *Supplement*, Robert Norman in his *New Attractive* – the same Robert Norman, skilled navigator and ingenious artificer, who first discovered the dip of the magnetic needle.[8]

Gilbert is not indiscriminate in his enthusiasm, however:

> Many others I pass by of purpose: Frenchmen, Germans, and Spaniards of recent time who in their writings, mostly composed in their vernacu-lar languages . . . seem to transmit from hand to hand, as it were, er-roneous teachings in every science and out of their own store now and again to add somewhat of error.[9]

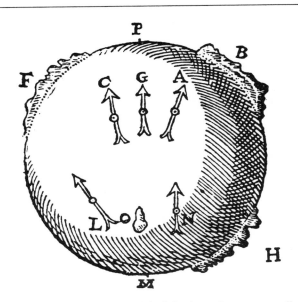

Figure 3.2 *William Gilbert's illustration of the behaviour of compass needles on the surface of a magnetic terrella, or artificial earth.*

Foreigners rely on the authority of others, which they promulgate uncritically, while Gilbert's good English authors simply record new ways of doing things or new instruments to aid the inquirer.

In the vein of predecessors that include Paracelsus, Gilbert notes that "men are deplorably ignorant with respect to natural things, and modern philosophers, as though dreaming in the darkness, must be aroused and taught the uses of things, the dealing with things; they must be made to quit the sort of learning that comes only from books, and that rests only on vain arguments from probability and upon conjectures."[10] One of the ironies of *De magnete* is in fact the extent to which Gilbert himself cites the views of older, frequently classical, authors, albeit often to claim that they are in error. Gilbert's book was not itself written for an unlearned readership, however, any more than had been Agricola's *De re metallica. De magnete* is in Latin, and at one point Gilbert even quotes a passage from Aristotle in the original Greek.[11] *De magnete* contains geometrical diagrams as well as naturalistic depictions of mathematical instruments and magnets used in experimental work, but it carries also pretensions to high culture. Gilbert co-opts the expertise of the navigator; he does not defer to it, nor does he present himself as one of their number, or indeed as any kind of mathematical practitioner. Instead, he claims to be in search of *causes*: Gilbert strives to be a *philosopher*.[12]

III Francis Bacon: philosophy, practical knowledge, and the place of antiquity

The strength of these sentiments in the later sixteenth century was, as we have seen, particularly evident in England. The most prominent, and much the most influential, example is found in the work of Francis Bacon, whose publications in the early seventeenth century have been credited with spurring a growth in empirical and utilitarian research in mid-century England and with promoting the foundation of the Royal Society of London in the 1660s.

Francis Bacon was born in 1561, the son of Sir Nicholas Bacon, a prominent courtier who became Keeper of the Great Seal (one of the major political offices of the time). Francis was trained in the law, and clearly aimed from the beginning at the same sort of career as his father. By the closing decades of the century, however, the long reign of Elizabeth I had led to something of a bottleneck in career paths for ambitious young courtiers. Although by the 1590s moving in government circles at court (and having held a seat in the House of Commons since 1584), Bacon was to find that higher office was not quickly attained, and that his executive power was restricted.

It was in these circumstances that Bacon first tried out his own plans for a renewed, practically-oriented natural philosophy in the service of the state. Rather like other Elizabethan promoters of operational knowledge, he conceived of the establishment of state institutions that would be dedicated to improvements in the crafts and trades. Not only did Bacon wish to see greater state attention paid to such knowledge, but he also began to develop theoretical ideas about the requisite cognitive structure of a practical, fruitful natural philosophy. Being so close to the centres of power, he naturally had ambitions to see his plans realized through direct governmental action. He lobbied unsuccessfully in Elizabeth's court during the final decade of the sixteenth century for the implementation of his design, which included a menagerie, botanical gardens, a research library, and a chemical laboratory. Despite the support of the Earl of Essex, one of the most powerful courtiers, Bacon's plan came to nothing. When James I succeeded Elizabeth in 1601, Bacon tried again, with no more success. It was at this point that he began to concentrate on the writing of manifestos, descriptions of his ideas intended to encourage the reforms for which he continued to hope.

The first of these texts to appear in print was an English composition on the *Advancement of Learning* (1605). It is a work that presents many of the basic arguments and rhetorical strategies that also feature in Bacon's later writings. The fullest expression of his views, however, is found in the

Novum organum ("New Organon") of 1620, with its hopeful dedication to King James. The work's very title indicates something of its character. Written in Latin (and hence accessible only to the well-educated), it purports to be a wholesale replacement for the complex apparatus of Aristotelian logic. The corpus of Aristotle's logical writings was traditionally known as the *Organon*, a Greek word meaning "tool" or "instrument." This was because logic was seen as an instrument to be used in all manner of specific subject-areas, restricted to none. In announcing his *new* organon, Bacon signalled his belief that the Aristotelian approach to logic pursued in the schools was a thoroughly inadequate instrument, particularly for the purpose of generating natural philosophical knowledge. His replacement was by contrast perfectly suited, he claimed, for precisely that goal.

Bacon's argumentative strategy was thoroughly radical. In challenging orthodox natural philosophy, he did not simply criticize the usual means for pursuing it. Instead, he advocated a reconceptualization of natural philosophy *itself*. He attacked Aristotelian natural philosophy, as well as many other alternatives, for being wrongly structured: he rejected the contemplative ideal for natural philosophy altogether. Instead, he held that natural philosophy, properly understood, should be directed towards achieving improvements in the well-being of humanity – what we would nowadays think of as technological advances. Mere tinkering with scholastic natural philosophy would therefore be of little use; the whole enterprise needed to be thought out anew. This theme is really the core of Book I of the *Novum organum*.

Bacon was careful, however, not to present himself as a straightforward foe of established, and especially ancient, authority. Indeed, one of Bacon's earlier publications had been a work of 1609 called *De sapientia veterum* (*On the Wisdom of the Ancients*), in which he had praised the philosophical insight of the so-called Presocratics, those Greek philosophers earlier in time than the great age of Socrates, Plato, and Aristotle (late fifth and fourth centuries BC). The Presocratics had the supreme advantage of being represented not by their own original writings but by the second-hand accounts of later Greek writers (first among them being Aristotle himself) and by brief quoted extracts from their works ("fragments") also found in later writers. This left a considerable degree of latitude for interpretation, permitting the ascription to them of a wide array of ideas and achievements that the evidence at least did not obviously forbid.[13] Bacon's criticisms of Aristotelian and Platonic doctrine were thus tempered by an avowed admiration for the achievements of even more ancient authorities.

Early in Book I of the *Novum organum*, Bacon took care to observe that "The honour of the ancient authors stands firm, and so does everyone's honour; we are not introducing a comparison of minds or talents but a comparison of ways."[14] The evidence of *De sapientia veterum* itself shows that such a remark should not, perhaps, be taken as a matter simply of deflecting criticism that he was not giving the ancients their due; Bacon, like other critics in the seventeenth century, frequently made a point of distinguishing between Aristotle himself and his latter-day self-styled followers. Bacon's idea was that of a bureaucratic administrator: progress would be made not by the fortuitous appearance of unusually capable individuals, but by the proper organization of collective effort. No doubt Aristotle had been clever, but that was not the point: "we are not taking the role of a judge," said Bacon, "but of a guide."[15]

Novelty was an important feature of Bacon's perspective, since it was new discoveries that he sought. His attitude towards the justificatory function of appeals to antiquity and towards humanist academic culture (to which he, like all others of his educational background, was inevitably indebted) was ambivalent in this regard too. Should novel ideas be presented as genuinely novel? In the *Novum organum*, Bacon in fact explicitly rejects the use of ancient authority to bolster his position. He notes: "it would not have been difficult for us to attribute our proposals either to the ancient centuries before the times of the Greeks [i.e., to the Presocratics and to ancient eastern sages] . . . or even (for part of it) to some of the Greeks themselves." But this would be imposture:

> We do not think that it is any more relevant to the present subject whether the discoveries to come were once known to the ancients . . . than it should matter to men whether the New World [i.e. America] is the famous island Atlantis which the ancient world knew. . . . For the discovery of things is to be taken from the light of nature, not recovered from the shadows of antiquity.[16]

Despite this confident pledge of allegiance to the cult of modernity instead of that of antiquity, standing in clear opposition to the humanist views that we have already seen, Bacon could not in practice avoid entirely the presentational and rhetorical techniques of his intellectual forebears. Scholars like Copernicus or Vesalius had set themselves up as opponents of the status quo by appealing to the precedent of an ancient world when things were better; in doing so, they had adopted the commonplace tactic of telling a tale of decline leading down to the present, when things were now to be set straight again. Bacon too tells a tale of decline so as to dispossess contemporary establishment philosophy of its authority. His tale takes a different form, however. Bacon again credits the Presocratic period as that "in which natural philosophy seemed to flourish most among the Greeks," but he does so in order to observe the brevity of its duration. Later on, he says, "after Socrates had brought philosophy down from heaven to earth, moral philosophy grew still stronger, and turned men's minds away from natural philosophy."[17] (Elsewhere he says that the works of the Presocratics were overwhelmed by "lighter works" that were chiefly concerned with pleasing the "taste of the crowd," and that "time (like a river) has brought down to us the lighter, more inflated works and sunk the solid and weightier.")[18] The Romans subsequently concentrated on moral philosophy and public affairs, while Christianity in the West had devoted its greatest rewards and its best minds to theology. It was no wonder, therefore, that natural philosophy had made such little progress. The lack of progress could be explained by a lack of application, and did not speak against the possibility of realizing Bacon's great vision of the benefits to be gained from a properly conducted – and socially supported – natural philosophy.[19] In effect, the capabilities of natural philosophy ought not to be judged by the achievements of its current practitioners, because history shows that we can do better.

Like a true revolutionary, Bacon refused to be held to account by the criteria of evaluation used by his opponents. Following the earlier remark about the untouched honour of the ancients, he explains that

> No judgement can rightly be made of our way (one must say frankly), nor of the discoveries made by it, by means of *anticipations* (i.e. the reasoning currently in use); for one must not require it to be approved by the judgement of the very thing which is itself being judged.[20]

Bacon's vision of natural philosophy, in contrast to the Aristotelian, saw it as an endeavour that would be productive of *works*; that is, of practical applications. This was so much the case that he spoke of this productiveness as not merely a *consequence* of proper natural philosophical knowledge, but as the very *criterion* of its truth. In a remark that expresses the point famously (if also, in its original Latin, somewhat ambiguously), Bacon said that "truth and usefulness are (in this kind) the very same things"; he goes on to say that "works themselves are of greater value as pledges of truth than for the benefits they bring to human life."[21] This is not to say, however, that Bacon therefore saw practical works as no more than means to the end of finding philosophical truth. Thus, in criticizing the inquiries of other philosophers in the *Novum organum*, he remarks at one point that "men do not cease to abstract nature until they reach potential and unformed matter,[22] nor again do they cease to dissect nature till they come to the atom. *Even if these things were true, they can do little to improve men's fortunes.*"[23]

Philosophy, for Bacon, was not an end in itself. Book I of the *Novum organum* is largely devoted to undermining the pretensions of existing philosophical schemes so as to clear the way for the establishment of his own approach, designed to supersede them all. But his main strategy is not one of analytical criticism, aimed at showing the ineffectiveness or ungroundedness of his rivals' arguments; instead, he concentrates on impugning their goals as unworthy. The fault of the Aristotelians lies above all in their misconstrual of the *purpose* of natural philosophy. To show contempt for practical knowledge, the sort that can provide humanity with a better life, is to act immorally. Bacon expresses this view using a Christian idiom that forms an important part of his entire position: Aristotle's unproductive philosophy is a dereliction of the Christian duty of charity towards others. Natural philosophy can in principle help people, and so it must be directed to that purpose. "The true and legitimate goal of the sciences is to endow human life with new discoveries and resources":[24]

> Just let man recover the right over nature which belongs to him by God's gift, and give it scope; right reason and sound religion will govern its use.[25]

The "right" of which Bacon speaks is something to be claimed, or asserted. Once the propriety of seeking power over nature has become accepted, then its pursuit becomes a matter of "right reason and sound religion": religion, because the goal will be knowledge for proper Christian purposes, and reason in the form of Bacon's new organon.

IV Knowledge and statecraft

Bacon presented his method, what he called a *via et ratio* ("way and procedure") for the making of knowledge, as being starkly opposed to the Aristotelian approach advocated in the academic world. Above all, Bacon criticized the Aristotelian fixation on the demonstrative syllogism.[26] A logical point underlay his condemnations. When a conclusion was drawn from its premises in syllogistic logic, that conclusion was about a particular. The major premise from which it was derived, however, was a *universal* statement. As an illustration, we may consider once again the classic syllogism:

All men are mortal	Major premise
Socrates is a man	Minor premise
Therefore Socrates is mortal.	Conclusion

Bacon's point, in essence, is that we are only prepared to accept the truth of the major premise because we already believe it to be true of all particular instances. Of these particular instances, Socrates's mortality is just one case. In general, therefore, our knowledge of the major premise is the *result* of our knowledge of a large number of individual instances like that of Socrates. The syllogism therefore argues backwards: in order to gain true knowledge, Bacon maintains that one must work back *from* individual instances *to* the universal knowledge-statement. The latter comes at the end of the knowledge-making process, therefore, not at the beginning (as it does in the syllogism). Thus, "[a]s the sciences in their present state are useless for the discovery of works, so logic in its present state is useless for the discovery of sciences."[27] Indeed, "the sciences we now have are no more than elegant arrangements of things previously discovered, not methods of discovery or pointers to new results."[28]

Bacon placed his faith in a particular form of "induction." By this word (or the Latin *inductio*), Bacon meant the creation of general truths ("axioms") concerning aspects of nature that would be analogous to the major premise of a deductive syllogism. Induction, moving in the opposite direction to that of deduction, would create the best-founded such truths, much superior to those currently accepted – for "[c]urrent logic is good for establishing and fixing errors (which are themselves based on common notions) rather than for inquiring into truth."[29] This induction was not, however, to be simply an accumulation of instances leading to an abstracted generalization: Bacon explicitly rejects "induction by enumeration" (a common ploy in the discipline of classical rhetoric) as childish. Rather than just piling up examples, and perhaps ignoring inconvenient exceptions, one should employ "rejections and exclusions," so as to end up with the truth by having eliminated all possibilities but one.[30] The result will, ideally, be a general statement ("axiom") built on experiential particulars, and having in addition a scope greater than that of the particulars from which it was derived. It would thus point the way to the discovery of new particulars.[31]

Bacon privileges the knowledge of the artisan in discussing the sources of natural-philosophical experience and their promise of leading to works. He extols the value of the relatively recent innovations in Europe of gunpowder, silk thread, the magnetic compass for navigation, and printing with moveable type, characterizing them all as new discoveries that had been stumbled upon by untutored, albeit practical, people. If these things could be found by chance, he argues, how much more might be expected from a disciplined, methodical inquiry.[32] He aimed, as he put it, at an experience "finally made literate."[33] This meant, in practice, the production of written lists of individual facts drawn from experience, which would be employed in the sorting process by which the higher axioms would be derived by elimination. These lists, or "tables," and their use are explained and discussed in Book II of the *Novum organum* together with illustrative examples.

The historian Julian Martin has characterized Bacon's approach to an active natural philosophy as that of a lawyer and civil administrator – such as Bacon, as we have seen, actually was. Towards the middle of the century, William Harvey (whose theory of the circulation of the blood had a great impact in the study of physiology and medicine) reportedly said of Bacon that he "wrote philosophy like a Lord Chancellor."[34] Usually taken as a dismissal of Bacon's philosophy, this remark nonetheless contains a great deal of truth, as Martin has shown. Bacon was involved in a project, at the beginning of the seventeenth century, intended to codify and systematize English law. This involved determining the precedents actually accepted by judges in the determination of court cases, as a means of reducing the common law to a definitive body of explicit statute law. The project called for the organization of cases into taxonomic categories, from which the legal principles underlying the decisions made in them could be abstracted. This, of course, is a bureaucratic piece of statecraft remarkably similar to Bacon's later programme for the reform of natural philosophy, a project that, like legal reform, Bacon put forward as one that would be in the interests of the state and of centralized control.

Bacon outlined in some detail his vision of the political organization of knowledge-making in the *New Atlantis*, which was published (in English) in 1626, the year of his death. The book presents a fabulous account of a mysterious island in the Pacific, unknown to Europeans, with its capital Bensalem. Above all, the island is a rationally governed state in which men concerned with the generation of useful knowledge play a central role. Such utopian visions were by no means unprecedented at this time; two seventeenth-century examples that preceded Bacon's, Johann Valentin Andreae's *Christianopolis* and Tommaso Campanella's *City of the Sun* (1619 and 1623), described ideal cities that, like Bensalem, instantiated a philosophical vision of the creation and transmission of knowledge about the natural world. The intellectual hub of Bacon's version was an institution called "Salomon's House." Bacon describes the number of personnel involved and their strictly segregated roles: people (all are men) to travel the world gathering facts, people to conduct experiments to generate new facts, people to extract from books candidate facts to be tested experimentally, and further up the hierarchy, men to consider all these experimental outcomes and direct the performance of new experiments. At the top of the tree were the Interpreters of Nature, men (three of them) who would take these solidly attested facts and use them to produce the axioms that were the crowning glory of Baconian inductive philosophy. In addition, there were others whose sole task was to draw conclusions from these axioms so as to yield specific practical benefits. Salomon's House, so constituted, was directed towards "the knowledge of Causes, and secret motions of things; and the enlarging of the bounds of Human Empire, to the effecting of all things possible."[35]

Bacon's concern to get at natural causes, the province of qualitative natural philosophy, gave him a correspondingly dull interest in mathematics – his conceptual categories, that is, betray a continuing debt to the philosophy of scholastic Aristotelianism. He spoke of mathematics in the *Novum organum*, remarking that it "should only give limits to natural philosophy, not generate or beget it."[36] Bacon wanted to know how things work, so as to be able to control them; that meant tangible causes, including the "secret motions of things." Although not a central element of his major discussion in the *Novum organum*, a theory of matter, to do with the behaviours of submicroscopic particles, their motions, sympathies and antipathies, played a significant role in Bacon's view of natural philosophy. Despite his avowed rejection of their work for its secretiveness and lack of concern for the public weal, Bacon's matter-theory in fact owed a great deal to the alchemists and magicians, as for example his use of such alchemical notions as that of the natural sympathies and antipathies to be found between the smallest parts of various substances. The ideal of the magus, harnessing the powers of the stars, which implicated a kind of knowledge that had practical ends, itself clearly coloured Bacon's own conception of the value of natural philosophy. Together with the increasing cultural value placed on artisanal knowledge, the field of magic (especially so-called natural magic, which made use of the hidden properties of natural things) provides an indispensable context for understanding the sources of Bacon's ideas in natural philosophy.

The real significance and consequence of Bacon's writings, however, relate centrally to his methodological opinions – rather than to the substantive content of his views on nature and its make-up. Bacon spoke of such exotica as "latent process" and "latent structure" in describing the hidden inner particulate and spiritual structure of kinds of matter,[37] but in practice those ideas acted as little more than physicalizations of his methodological precept that knowing *what* a thing is and knowing *how to produce* it are essentially identical. Creating gold, for example, by superinducing upon a piece of matter the appropriate qualities of gold – to make the matter yellow in colour, of a particular density and malleability, and so forth – was a process that he understood in conventionally "mechanical" terms.[38] That is, these were properties to be given to a body by application of the techniques of a craftsman, or "mechanic"; hammering, heating, sifting – any of the sorts of operations that can be done on *pieces of matter*. So the effective underlying structure of matter was accommodated to that model: matter is made up of parts that can be reshaped, rearranged, beaten, jostled around by heating, and suchlike.[39] As Chapter 7 below will show, Bacon's stress on first-hand experience and experiment, together with a high evaluation of utility, was used to promote precisely the kind of pragmatic corpuscular mechanism that is so typical of the early Royal Society later in the century.

Chapter Four
Mathematics Challenges Philosophy: Galileo, Kepler, and the Surveyors

I Natural philosophy – the only game in town?

Bacon's notion of an operational natural philosophy took its lead from the kinds of natural philosophy taught in the schools. Bacon attempted a radical reformation of natural philosophy, but it was still a reformation rather than a completely different enterprise. This fact might suggest that the available scope for rethinking the study of nature was severely restricted – as indeed it was. But natural philosophy was not the only model provided by learned culture for the study of nature. There were other relevant areas of inquiry too, areas that could be turned to account by people dissatisfied by (or uninterested in) the enterprise of the physicists.

Recall that Aristotelian physics aimed at understanding qualitative processes. Quantities were at best peripheral to it, because they failed to speak of the essences of things – of what *kinds* of things they were. Measurements, whether of dimensions or of numbers, were purely descriptive, while the natural philosopher's job was defined by its attempt to *explain*, not merely describe.

During the sixteenth century, certain Aristotelian philosophers had denigrated the mathematical enterprise on precisely these grounds. Scholars like Alessandro Piccolomini, and prominent natural philosophers like Benito Pereira, published critiques of mathematics that contrasted it unfavourably with physics. Mathematics, they said, did not demonstrate its conclusions through *causes*. This disqualified mathematical proofs from being scientific in Aristotle's sense, because Aristotle had specified that true scientific demonstration always proceeded through the identification of a relevant explanatory cause for its conclusion. Such causes, falling under one of Aristotle's four categories of formal, final, efficient, and material, were what made a proof into a piece of science.[1] None of these kinds of cause was utilized in mathematics, its critics claimed, and hence mathematics was not a scientific discipline. Indeed, the most damning short-

65

coming of all was mathematics' failure to speak of *formal* causes, that is, explanatory causes that relied on specifying the *kind* of thing that was involved. In other words, mathematics did not get at the true *natures* of its objects, and was restricted to discussing only superficial quantitative properties (in Aristotelian terminology, quantitative *accidents* unrevealing of a thing's nature, or *essence*).

Needless to say, there were contemporary mathematicians who resented such assertions. They wished to portray their own discipline as a "science" because that was the highest grade of knowledge; they did not want second-class status behind the physicists. Accordingly, several mathematical writers in the later sixteenth century and the early seventeenth century produced counter-arguments to establish, against the natural philosophers, that mathematical proofs were indeed causal and properly scientific. Foremost among them were mathematicians belonging to the Catholic religious order called the Society of Jesus – the Jesuits.

During the second half of the sixteenth century the Jesuits (founded by Ignatius Loyola in 1540) became the foremost teaching order in the Catholic world. Their colleges quickly sprang up all over Europe, with a reputation for excellence that was second to none. The education that the Jesuit colleges offered was comparable to the arts education available at universities. Apart from the explicitly religious aspects, which underlay the whole, Jesuit education thus consisted of a great deal of humanist training in ancient languages and literature, as well as education in the traditional scholastic subjects based on the texts of Aristotle – physics, metaphysics, and ethics, together with the subjects of the quadrivium, that is, mathematics.[2] The Jesuit mathematicians were frequently different people from those who taught natural philosophy, and some of them objected to the belittling characterizations of their specialty found even in the writings of their own philosophical brothers, such as Pereira. The earliest concerted defence came from the leading Jesuit mathematician of the late sixteenth century, Christoph Clavius, professor of mathematics at the Jesuits' flagship college in Rome, the Collegio Romano. Clavius explicitly rejected the claims of the philosophers concerning mathematics, and pointed out the pedagogical harm that could be caused by their teachings on the subject. There were those, he complained in the 1580s, who told their pupils that "mathematical sciences are not sciences, do not have demonstrations, abstract from being and the good, etc.".[3] Clavius wanted the teachers of mathematics to be accorded as much respect as the teachers of natural philosophy and metaphysics, and the scurrilous charges against mathematical knowledge hindered this goal. As regards substantive responses to the hated arguments, Clavius himself was less effective, although he established a position in support of mathematics that was subsequently widely imitated by other Jesuit mathematicians. He relied especially on Aristotle's own discussions, pointing out that Aristotle had included mathematics as an integral part of philosophy alongside natural philosophy, thereby implying that it had an equivalent cognitive status, and that Aristotle had described the mixed mathematical disciplines (astronomy, music, and so on) as being "subordinate sciences"; that is, sciences that relied on results borrowed from other higher sciences – meaning arithmetic and geometry. There could thus be no doubt that Aristotle regarded mathematics as truly scientific.

Later Jesuit mathematical writers supplemented Clavius's appeals for fair play with philosophically-based refutations of the anti-mathematical arguments. A former student of Clavius, Giuseppe Biancani, in a work of 1615, wrote at some length on the question, denying the view that mathematical demonstrations did not employ causal proofs and that mathematical objects (geometrical figures or numbers) lacked true essences – in effect, that they were not real things. Biancani says that, on the contrary, geometry defines its objects in such a way as to express their essences. He means that a triangle, for example, is a figure composed of three right lines in the same plane that intersect one another to yield three internal angles – that is what a triangle *is*. Similarly, geometrical figures have their own matter (the subject of material-cause explanations), in this case *quantity*. Using such arguments, Biancani attempted to refute the philosophical critics of mathematics, while also following Clavius in claiming a certain *superiority* for mathematical demonstrations over those of natural philosophy. This superiority flowed from the generally accepted certainty of mathematical proofs, which by common consent exceeded that of other kinds of philosophical argument.

Thanks initially to Clavius, these sorts of arguments were well known, especially among Jesuit mathematicians, in the early seventeenth century. They served as a means of increasing the confidence of mathematicians that their sciences were not only on a par with natural philosophy but were perhaps in some ways even better at making reliable knowledge of nature. One such mathematician was an Italian friend of Clavius, Galileo Galilei.

II Galileo the mathematical philosopher

Galileo was born at Pisa, the second city of the Grand Duchy of Tuscany in northern Italy, in 1564. He was the son of a musician, Vincenzo Galilei, who was from Tuscany's capital city, Florence, and Giulia Ammannati, and the family held minor noble status derived from its Florentine forebears. Galileo attended the University of Pisa to study medicine, but his lack of vocation conspired with his aptitude for mathematics to cause him to leave in 1585; he subsequently returned to the university in 1589 to take up a chair in mathematics. The chair had been secured on the strength of personal recommendations from established mathematicians, especially Guidobaldo dal Monte (Galileo had also met Clavius by this time, on a visit to Rome in 1587).[4]

Much of Galileo's subsequent career must be explained by reference to his aggressive and ambitious personality. His approach, however, and the values that he expressed, were not idiosyncratic, but can be understood as part of the outlook of a university mathematician of his time and place. Although other people in similar positions failed to acquire Galileo's fame, Galileo did what many of them would no doubt have liked to achieve – he stood up to the higher-paid, more prestigious natural philosophers and refused to concede to their expertise.

The earliest example of this dates from around 1590, during Galileo's professorship at Pisa. An early manuscript treatise surviving from that period, usually known as *De motu* ("On Motion," composed in Latin), signals by its very title that Galileo intends to take on the despised Aristotelian physicists. Motion, as an example of change, was a central topic of Aristotelian physics. The natural philosopher spoke of motion so as to explain why things moved, and one of the typical kinds of such explanations invoked an appropriate final cause. In particular, to explain the free fall of a heavy body, Aristotle had described it as a *natural* motion, since it is in the nature of heavy bodies to fall when unimpeded. But *why* do they fall? Aristotle decided that they fell because they were seeking their proper place at the centre of the universe. Fall thus appeared as a *process* of travel, wherein the moving body set off from its starting place in an endeavour to reach its goal. That goal, the centre of the universe, coincided in Aristotle's cosmos with the centre of the earth – because the earth is simply the accretion of all heavy bodies bunched together around their natural place, towards which they strive.

One of the Aristotelian rules governing fall that emerged from this conceptualization was that the heavier a body, the faster it falls. Weight expressed the motive tendency of the body, so if weight increased, so too should the speed of descent. A body that weighs twice as much as another ought therefore to descend twice as quickly as the lighter body. Galileo, in *De motu*, argues that this familiar Aristotelian claim is false, and he provides a number of arguments intended to show it. One, for example, imagines two independently falling bodies becoming linked together by a piece of cord as they fall. Becoming connected, they should now constitute a single aggregate body. Such a body, being heavier than either of its original components, would, according to Aristotelian doctrine, fall more rapidly than either one. And yet, Galileo urges, it is not conceivable that the two pieces would suddenly speed up as soon as the cord linked them.

Galileo's strategy becomes clearer when he calls on the precedent of the ancient mathematician Archimedes to aid him.[5] In *On Floating Bodies*, Archimedes considered the relationship between the specific gravity (or density) of a body and that of the medium in which it was immersed. He used this relationship to determine whether the body should float or sink: if the body was denser than the medium, it sank; if less dense, it floated. Galileo takes the same approach in his own discussion of falling bodies –

in effect, he treats falling bodies as if they were all *sinking* in a common medium, the air, and compares their rates of fall by comparing their specific gravities in relation to air's.

Galileo, notably, does *not* ask the question "*why* do heavy bodies fall?" That would have been a natural philosopher's question. Galileo, the mathematician, asks only how fast they fall, and what the relationship is between their densities and that of the medium; like Archimedes, Galileo does not ask what weight *is*. Against Aristotle, he concludes, first of all, that two bodies of differing weights – say, differently sized iron balls – will nonetheless fall at identical speeds. The speeds are a function of the balls' specific gravities in a common medium, air; since both balls are made of the same material, solid iron, their speeds too are the same.

In 1591 Galileo left the university at Pisa to take up a similar, although rather more illustrious, professorship at the great thirteenth-century university of Padua. The city of Padua, in north-eastern Italy, was at this time a part of the independent republic of Venice, and Galileo's academic position fell under the control of the Venetian senate. For nearly two decades Galileo remained at Padua, lecturing on mathematical subjects and engaging in occasional controversies with Aristotelian philosophers there. He supplemented his income by making and selling mathematical instruments designed for surveying work, an activity that was a common feature of practical mathematical pursuits at the time.[6] By 1609 he had developed to a high degree his work on the motion of heavy bodies, including the famous doctrines of the uniform acceleration of freely falling bodies and the parabolic paths of projectiles. This work, however, was not to be published until 1638, in his *Discorsi* ("Discourses and Demonstrations Concerning Two New Sciences," often referred to in English as the *Two New Sciences*).[7] His aversion, as a mathematician, to the natural philosophy of his Aristotelian colleagues continued to motivate him, and probably contributed to his readiness, from the 1590s onwards, to entertain the unorthodox doctrines of another mathematician, Nicolaus Copernicus.

Galileo's interest in Copernicanism existed from at least 1597, when he mentions Copernicus in two letters. One of these letters was sent to the great astronomer Johannes Kepler, acknowledging receipt of the latter's Copernican book *Mysterium cosmographicum* ("Cosmographical Mystery") of 1596; Galileo, famously, claims to Kepler that he too was a Copernican, and had been "for many years."[8] It was not until the first decade of the seventeenth century, however, that Galileo took up astronomical and cosmological issues in a serious way, especially from 1609 onwards when he began to use a telescope to make astronomical observations.[9] Copernicanism seems to have appealed to Galileo above all because it was a useful tool for attacking the Aristotelian physicists. First, it advocated the acceptance of a sun-centred universe, which would tear to shreds the physical world-picture on which the entire Aristotelian system was based. If the earth were no longer at the centre of the universe, for example, the fall of heavy bodies (and the rise of light bodies) could no longer be explained by their desire to reach a destination defined in terms of the centre of the universe, because the latter would no longer coincide with the earth's centre.[10] Secondly, the chief arguments in favour of Copernicanism were astronomical rather than cosmological: that is, they were the arguments of a mathematician, concerned with reducing the apparent motions of the heavens to order, rather than those of the physicist, concerned with the nature of the heavens and the explanation of their movements. At the same time, Copernicus and a few followers of his doctrine, such as Kepler, had embraced the cosmological inferences that they nonetheless dared to draw from the new astronomical system.[11]

Galileo therefore attempted to use Copernican astronomy as a mathematician's means of subverting Aristotelian cosmology. He trampled on the usual demarcation between physics and mathematics by stressing that the natural philosopher had to take into account the discoveries of the mathematical astronomer, since the latter concretely affected the content of the natural philosopher's theorizing – the astronomer told the physicist what the phenomena were that required explanation. In his *Letters on Sunspots* (1613), Galileo made this point strongly in arguing for the presence of variable blemishes on the sun's surface. The Aristotelian heavens were held to be perfect and substantively unchanging; all they did was to wheel around eternally, exhibiting no generation of new things or passing away of old. The marks first seen on the face of the sun by Galileo and others in 1611 did not appear to show the permanence and cyclicity characteristic of celestial bodies, and Galileo took the opportunity to argue that they were, in fact, dark blemishes that appeared, changed, and disappeared irregularly on the surface of the sun. It was important to the argument that the spots be located precisely on the sun's surface itself. The Jesuit Christoph Scheiner, Galileo's main rival for the glory of their discovery, at first thought that the spots were actually composed of small bodies akin to moons, which orbited around the sun in swarms so numerous as to elude, thus far, reduction to proper order. Accordingly, Galileo presented careful, geometrically couched observational reasoning to show, first of all, that there was an apparent shrinkage of the spots' width as they moved across the face of the sun from its centre towards the limb (and corresponding widening as they appeared from the other limb and approached the centre); and secondly, that this effect, interpreted as foreshortening when the spots were seen near the edges of the sun's disc, was consistent with their having a location on the very surface of the sun itself. The precise appearances, he argued, would be noticeably different if these necessarily flat patches were any distance above the sun.[12]

Galileo's argument leads to the following point: if it is established that the sun's surface is blemished by dark patches that manifestly appear from nothing and ultimately vanish, then it becomes undeniable that there is, contrary to Aristotelian doctrine, generation and corruption in the heavens. Galileo has moved from a "mathematical" explication of the

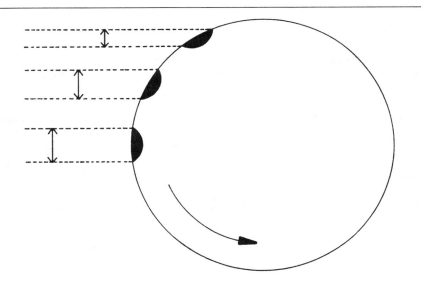

Figure 4.1 *Galileo's reasoning concerning the foreshortening of sunspots as they approach the sun's limb, to show that they are on the sun's surface.*

external properties of things (here, the apparent size, shape, and motion of the sunspots) to a properly *physical* conclusion about the matter of the heavens.

As he explained elsewhere in his published contributions to the debate with Scheiner, the true essences of things as distant as the celestial bodies cannot be determined by the senses, and indeed the same should be understood also of bodies near at hand: "I know no more about the true essences of earth or fire than about those of the moon or sun, for that knowledge is withheld from us, and is not to be understood until we reach the state of blessedness."[13] Hence all that remains to us is knowledge of those manifest properties which *are* accessible to the senses.

> Hence I should infer that although it may be in vain to seek to determine the true substance of the sunspots, still it does not follow that we cannot know some properties of them, such as their location, motion, shape, size, opacity, mutability, generation, and dissolution. These in turn may become the means by which we shall be able to philosophize better about other and more controversial qualities of natural substances.[14]

Not only could the manifest (and measurable) properties of bodies be known, but such knowledge would enable better philosophizing. The work of the mathematician, that is, could guide that of the physicist.

III The rising status and cognitive ambitions of the mathematical sciences: Galileo and Kepler

Galileo sometimes used the self-descriptive label "philosophical astrono-mer"[15] to represent the kind of work that he purported to be achieving in his work on sunspots and on the Copernican world-system. There is a hint of continuing deference to the category of natural philosopher, if not to natural philosophers themselves, in the way he liked to characterize himself. While negotiating with the Tuscan court in 1610 over the terms of his new service to the Medici (see Chapter 6, section II, below), Galileo insisted that his official title be that of court "philosopher and mathematician." It was common for a princely court to retain a mathematician (Tycho Brahe and Kepler both played that role), but this was clearly insufficient for Galileo. He wanted to be recognized also, and perhaps first, as a philosopher, someone who had things to say about the nature, not just the disposition, of the universe.

The Jesuit Biancani's arguments for the full causal character of math-ematical demonstration expressed very much the same sentiment. In Biancani's case, however, there was no real attempt (Clavius's paean to the peculiar certainty of mathematics notwithstanding) to set up the techniques of mathematicians as potentally superior alternatives to those of the physi-cists. The Jesuit mathematicians' goal seems to have been one of achieving parity with their natural-philosophical colleagues; Galileo's goal was to reform natural philosophy itself, so that it would be recognized as a disci-pline for mathematicians. Either way, such promotion of mathematical sciences as exemplary ways of learning about the natural world typifies a widespread movement in the first half of the seventeenth century. It was a movement that began to be recognizable through its gradual adoption of an identifying label: "physico-mathematics."

The value of this label sprang from its imprecision. It served to unite the notion of the physical with that of the mathematical, but the nature of the juxtaposition was ambiguous. It apparently designated a kind of math-ematics (in the broad contemporary understanding of that word) that was in some way of physical relevance. There were older, pre-existent terms for what looks like the same thing, as we have seen in Chapter 1, section II: "mixed mathematics" was perhaps the most common. And yet there seems to have been a felt need for the new term. Why?

This is where Galileo is such a useful figure. His endeavours help us to understand what the spread of "physico-mathematics" meant to those who eagerly adopted the term. Galileo's polemics and propaganda set into relief, perhaps in exaggerated form, those issues the debating of which form the core of what we can call the Scientific Revolution. These issues con-cerned the question of the proper character of natural philosophy: what should it be about, how should it be pursued, and why? Chapter 3 con-sidered the attempts of people like Francis Bacon to reform notions of what the purpose of natural philosophy should be. In arguing that it ought to be directed towards practical utility, Bacon at the same time effectively altered the ways in which it should be conducted, as well as how its knowledge-claims should be constituted and presented (his new definition of "forms"). The endeavours of the mathematicians, while different in focus and scope, acted in concert with this new stress on knowledge for practical use to promote a view of natural philosophy that emphasized the operational. In doing so, they came close to rejecting natural philosophy in its old sense in favour of an entirely different enterprise, simply applying to it an old name borrowed from the rejected discipline.

The case of Galileo illustrates how this complete break in fact failed to take place. He, in common with users of the contemporary term "physico-mathematics," retained a claim to the label of natural philosopher. The properties that he and other mathematicians wished to attribute to mathematical knowledge, properties that they resented the physicists for denying to it, were lifted from natural philosophy itself. Mathematicians did not simply declare the virtues of the mathematical sciences in isolation from those of physics; the relative status of the two disciplinary areas meant that mathematicians would still have been left – however certain their demonstrations – in command of what most others saw as an inferior kind of knowledge. In this regard the mathematicians resembled the craftsmen. The change in values expressed by Bacon involved the investing of practical, artisanal knowledge with a higher social status. It had been (and to a considerable extent continued to be) associated with low-status work – manual labour. Bacon in particular argued for a higher evaluation of utility by claiming its importance for the state, as well as through moral and religious arguments that associated it with Christian charity. And yet he wanted this newly-upgraded practical knowledge to receive the prestige already possessed by natural philosophy. His solution was to argue as if "natural philosophy" were a category much broader in scope than usually admitted by academics, one that *included* practical knowledge; he then chased out purely contemplative knowledge by criticizing the goals of the latter, thus leaving the field to his own proposed endeavour.

Similarly, Galileo and other mathematicians rejected the disciplinary boundary between natural philosophy and mathematics by arguing that mathematics was crucially important in drawing legitimate *physical* conclusions. In effect, the label "physico-mathematics" served to signal that the mathematicians' own expertise would not thereby be subsumed to that of the natural philosophers. Instead, the cuckoo's egg of physico-mathematics would (if Galileo had his way) serve to expel most of the original occupants of the natural-philosophical nest, so as to leave the mathematicians in the position formerly occupied by the physicists. In both this and the previous case, the established category "natural philosophy" was a valuable resource for those who wanted to raise the status of their own favoured kind of knowledge.

Another important advocate of the central place of mathematics in natural philosophy was the Copernican astronomer Johannes Kepler. Kepler's approach to astronomy was, like any astronomer of the time, fundamentally mathematical. But he went much further in his promotion of mathematics than most of his colleagues: for Kepler, the mathematics that structured astronomical theory was the very mathematics that underlay the structure of the universe itself. Thus, in his work as a mathematical astronomer, Kepler at the same time endeavoured to create a mathematical *physics*. For Kepler, the universe is properly intelligible in mathematical terms; it is mathematics, especially geometry, which allows insight into the mind of God, the Creator, and hence into the deepest realms of natural philosophy. In one of his last publications, a work of 1618 called *Epitome astronomiae Copernicanae* ("Epitome of Copernican Astronomy"), Kepler describes his own special field as a *part* of physics:

> What is the relation between this science [astronomy] and others? 1. It is a part of physics, because it seeks the causes of things and natural occurrences, because the motion of the heavenly bodies is amongst its subjects, and because one of its purposes is to inquire into the form of the structure of the universe and its parts. . . . To this end, [the astronomer] directs all his opinions, both by geometrical and by physical arguments, so that truly he places before the eyes an authentic form and disposition or furnishing of the whole universe.[16]

Kepler put these principles into effect in his restructuring of Copernican astronomy. As a student at the Lutheran university in the German town of Tübingen, he had become convinced of the truth of the new Copernican cosmology from his teacher in astronomy, Michael Mästlin. Belief in the literal truth of the Copernican system, as opposed to a recognition of the value of Copernicus's *De revolutionibus* in the practical computational work of mathematical astronomy, was not widespread among astronomers at this time, and Kepler's early guidance by one of the exceptions to this rule is therefore noteworthy. Kepler's metaphysical and theological predilections expressed themselves in relation to Copernican astronomy in his first publication, the *Mysterium cosmographicum* ("Cosmographical Mystery") of 1596, when Kepler was working as a school teacher in Austria. The most noteworthy feature of the work is its presentation of Kepler's proud discovery of a relationship between the dimensions of the planetary orbits (calculated according to the Copernican system) and certain interrelationships among the so-called "perfect" or "Platonic" or "regular" solids.

The latter were solid figures that had been demonstrated by Euclid to be restricted to precisely five in number. They were solids that are contained by identical facets which are themselves regular polygons, such as equilateral triangles, squares, or pentagons. The five solids, as Euclid had shown, were the tetrahedron, the cube, the octahedron, the dodecahedron,

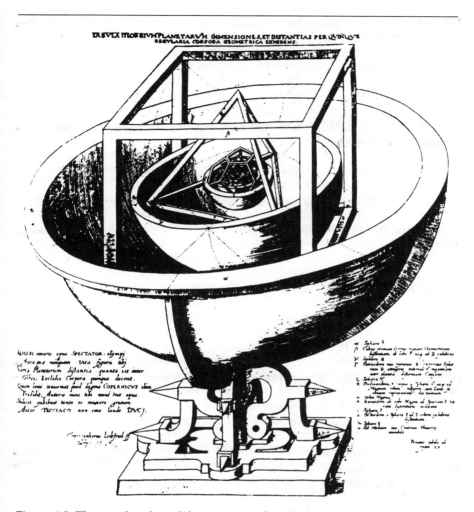

Figure 4.2 *The nested perfect solids structuring the universe, from Kepler's* Mysterium cosmographicum.

and the icosahedron, of four, six, eight, twelve, and twenty faces respectively. The fact that these five solids were unique of their kind implied to Kepler that they represented something profound about the nature of space and of the geometrical principles on the basis of which God had created the universe. In the *Mysterium cosmographicum*, he shows that (imaginary) spheres used to represent the relative sizes of the various Copernican planetary orbits around the sun are separated by various distances that closely accommodate the perfect solids as spacers between the spheres. Using available data, Kepler was able to show that the sizes of the planetary orbits closely fit the sizes allowed by the intercalated solids, to within an error of around five per cent. In 1600 he joined Tycho Brahe in Prague so as to gain access to Tycho's famed data on planetary motions, which Kepler hoped would enable him to reduce the error still further. Furthermore, Kepler's model accounted for there being, in a Copernican universe, precisely six planets – the number that could be adequately spaced by five intervening solids.

Kepler was enormously proud of this result, which he believed brought him nearer to an intimate understanding of the structure of God's Creation. The rôle of geometry in his argumentation was fundamental: geometry was not simply a tool for calculating dimensions and motions in astronomy; it was capable of providing *explanations* of why things in the world are as they are. The geometry of the five perfect solids serves not only to *describe* the number of the planets and their distances from the sun, but to make *sense* of those facts. Kepler believed in a fundamentally mathematical constitution to the universe, in the sense that mathematical intelligibility of the kind provided by the perfect solids accounted for *why* certain things are as they are. The nature of such an explanation is not, in the present case, one that provides mathematical, demonstrative *necessity* to the things that it explains (as with showing, as Euclid does, *why* the base angles of an isosceles triangle are equal to one another); but it does show, Kepler believed, what was in God's mind when He chose to create things in the way that He did. In many respects, in fact, Kepler's entire astronomical career was one directed towards gaining an understanding of God's mind, of coming closer to God through the medium of astronomical study. This was natural philosophy in its starkest, most theocentric form.

Kepler's major work was the *Astronomia nova* of 1609. It was the published result of a project that he had originally undertaken at the behest of Tycho, to determine a satisfactory astronomical model for the motion of Mars. Mars had always been a planet whose motion was particularly troublesome to model with exactness, and since Tycho's great observational project had been designed as the foundation for much more accurate planetary models, the continuing recalcitrance of Mars was a source of especial concern to him. Tycho was particularly interested in having Kepler solve the difficulties in terms of Tycho's own favoured cosmological system, a kind of compromise between Ptolemy and Copernicus that he had first published in a book of 1588. This scheme had the moon and sun in orbit around a central, stationary earth, but with the planets orbiting that moving sun. The resultant relative motions thus remained the same as in Copernicus's system (disregarding the issue of the fixed stars), with Copernicus's annual orbit of the earth around the sun being exactly mirrored in the annual orbit of the sun around the earth. Kepler responded to the challenge by producing models that could be expressed in Ptolemaïc, Copernican, or Tychonic terms (simply by shifting reference-frames). But, for Kepler, the Copernican remained the true account.

Several years of intensive work by Kepler resulted in an achievement that was remarkable in several ways. First, Kepler produced a model for the motion of Mars of unparalleled accuracy, both as determined by comparison with Tycho's observations and as confirmed over time by its predictions. Second, in doing so, he had come to abandon the classical Greek astronomical requirement, followed proudly by Copernicus as well as by Tycho himself, that the component motions used in creating astronomical models each be a uniform motion around a circle. Third, Kepler developed his new laws governing planetary motion on a basis that involved speculation about the *physical* causes that brought about that motion.

His new planetary orbits around the sun took the form of ellipses, with one focus of each ellipse located on the sun itself. He knew the geometry of the ellipse, one of the conic sections, from the treatise on conic sections written by the Greek astronomer and mathematician Apollonius of Perga, and Kepler's desire to find mathematics written in the fabric of the universe was thoroughly satisfied by this result, even though it meant abandoning circles. Furthermore, his elliptical orbits were traversed by the planets (including the earth) in such a way that the space swept out by the line joining the planet to the sun was uniform – equal areas swept out in equal times.

Equally importantly for Kepler, however, he had achieved these results in continual dialogue with ideas on the causes of planetary motions. These included the idea of a motive force emanating from the sun that drove the planets around in their orbits, together with an idea about a kind of magnetic attraction and repulsion between the sun and the two poles of each planet that served to explain why planetary orbits were not perfectly circular. Making explicit reference to William Gilbert, Kepler used his notion of the earth as a giant magnet to explain why planets successively approach and depart from the sun in the course of their elliptical orbits. The celestial spheres were gone (Tycho had already rejected them); Kepler's planets moved independently through space.

Kepler's views on the place of mathematics in understanding the physical world were thus more directly related to a purely philosophical, as opposed to practical, conception of natural knowledge than were Galileo's. The very nature of the mixed mathematical sciences, however, was such as to encourage, even in Kepler, a concern with some operational criteria of knowledge. The instrumental function of optics in assisting astronomical investigations was a major part of his justification for publishing *Ad Vitellionem paralipomena quibus astronomiae pars optica traditur* ("Additions to Witelo, in which the Optical Part of Astronomy is Treated"), in 1604.[17] Kepler considers the imperfection of sciences such as astronomy and optics, as compared to the demonstrative ideal of geometry, but argues

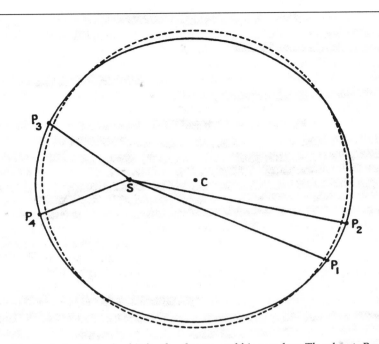

Figure 4.3 *Kepler's elliptical orbit for the planets, and his area-law. The planet, P, pursues its elliptical path with the sun, S, at one focus. The line joining the planet to the sun sweeps out equal areas in equal times, so that the distance traversed by the planet when nearer to the sun (P_3–P_4) is greater than that traversed when farther from the sun (P_1–P_2). From Marie Boas,* The Scientific Renaissance 1450–1630 *(New York: Harper and Brothers, 1962),* © *1962 by Marie Boas.*

that optical theorems should be sufficient to satisfy an astronomer's needs.[18]

IV Knowing, doing, and mathematics

Mathematics was itself traditionally related to practical endeavours such as land-surveying or the building of fortifications. Both fell under the heading of "mixed mathematics," along with such others as astronomy and mechanics. The latter too were of great practical importance. Astronomy had been valued in Latin Europe since the Middle Ages for its use in marine navigation and in astrology, a practical art much used in learned medieval medicine. Mechanics concerned machines themselves (such as wind or water mills), but more especially discussed the classical domain of the so-called simple machines, which considered certain devices and techniques

(such as levers or pulleys) that made work easier. The practical and artisanal associations of many of the mathematical sciences were thus very
hard to miss.

During the second half of the sixteenth century, mathematicians, especially in England, had begun to make strong claims for their discipline that
revolved around its practical dimensions rather than focusing on the more
philosophical justifications preferred by increasing numbers of bookish
mathematicians. In 1570 there appeared a new translation into English of
Euclid's *Elements*, bearing a preface written by John Dee of Mortlake. He
used this opportunity to praise the branches of mathematics for their
usefulness "in the Common lyfe and trade of men," as witnessed by the
practices of many and diverse occupations.[19] Dee had himself already had
dealings with one such endeavour, navigation; the interrelated concerns
of navigation (including in this period increasing interest in the magnetic
compass and terrestrial magnetism) and of cartography were important,
and unassailably mathematical, subjects of books by a number of English
authors in the decades around 1600, such as Robert Recorde, Thomas
Digges, and Edward Wright. Most such authors wrote in English rather
than Latin, and presented themselves as men of practical rather than contemplative bent. Typical examples of the genre include works on surveying techniques, the demand for which seems to have grown during the
second half of the sixteenth century, in concert with the increasing enclo-
sure of formerly common land and the surveying of church lands now
seized by the Crown following the English Reformation.

Mathematics thus had, besides its association with learned classical treatises and the niceties of formal demonstration, a practical, computational
image somewhat at odds with the academic, philosophical discipline promoted by scholars such as Clavius. At the same time, its leaning towards
practicality enabled it to appeal to the same sensibilities that Bacon's propaganda exploited. The kind of knowledge that mathematical practices
tended to promote was not simply utilitarian, however: its elevation to
philosophical importance by such as Galileo implied a revaluing of mathematical characteristics as being peculiarly important to true understanding of nature.

Chapter Five
Mechanism: Descartes Builds
a Universe

I A world to fit the knower

It was one thing to hold aloft particular ideals, operational or mathematical, concerning the sort of knowledge about the world that was desirable. But was the world itself the right *kind* of world for providing that knowledge?

Francis Bacon, as we saw in Chapter 3, does not seem to have worried overmuch on this point: when he dismissed questions about the divisibility of matter, he wrote that we need not worry about whether atoms are the ultimate constituents of matter or not, since "[e]ven if these things were true, they can do little to improve men's fortunes."[1] Bacon's vision of knowledge allowed some truths about nature to remain unknown, to form no part of his natural philosophy; he did not worry that such unanswered questions might compromise the *useful* answers one might acquire to other, addressable questions. But others had less eccentric conceptions of natural philosophy, conceptions that were not so single-mindedly tied to operational criteria. For such people, it did matter whether nature in itself was fully captured in their accounts of it. If the accounts were incomplete, so too was the natural philosophical enterprise – not only incomplete, but also potentially flawed, since who knew what unknown causes might be involved in subverting the effects of known ones?

For those who subscribed to the mathematical and operational ideal of natural knowledge, therefore, there were two main alternatives to pursue. One was to behave, like Bacon, in a pragmatic fashion, being satisfied with what works and leaving aside useless questions. The other was to put forward a view of the natural world that would consist of precisely those ingredients that a mathematical-operational form of knowledge was capable of discussing, *and no more*. The most successful and influential philosopher to adopt this second alternative, and to attempt to build a universe to suit his mathematical ideal of nature, was the Frenchman René Descartes.

Descartes was born in 1596, and was educated at the élite Jesuit college at La Flèche in northern France. After he left the college in 1614, Descartes acquired a law degree at Poitiers. By 1618 he had joined the Dutch army of Prince Maurice of Nassau, as a mercenary. This was a thoroughly unremarkable career development for a moderately well-to-do young gentleman in this period; it gave Descartes, as he later related, an opportunity to see a bit of the world as well as to mix with new people.[2]

One of the new people that he met in the Netherlands at this time was a schoolmaster named Isaac Beeckman. Beeckman is especially well-known to historians for his surviving *Journal*, which records many of his extraordinary ideas. The *Journal* also talks about his acquaintance with Descartes and the latter's intellectual virtues. In particular, it records Descartes's interest in Beeckman's own concern with micro-level mechanical explanations of natural phenomena. Beeckman attempted to develop hypothetical accounts of the causes behind various physical phenomena, accounts which were rooted in the idea of matter as being composed of minuscule bodies, or corpuscles. The shapes, sizes, and movements of the corpuscles were used as the exclusive causes of macroscopic, visible phenomena. For example, Beeckman wanted to explain magnetic attraction in terms of tiny corpuscles emitted from the magnet and impelling pieces of iron towards it by mechanical impact. This form of corpuscularianism, associated for Beeckman with the classical doctrine of atoms, went along with his pronounced interest in the traditional mathematical sciences. These included especially hydrostatics as well as such questions as the acceleration of freely-falling heavy bodies. Beeckman tended to elevate such sciences to a privileged status in attempting to understand the physical world, and was one of the first promoters of the term "physico-mathematics."[3] His liking for such explanations, including the more speculative corpuscular explanations, was evidently due to the fact that, for him, they "put sensible things as it were before the imagination."[4] In other words, much like Francis Bacon, he wanted physical explanations to be couched in the terms of practical mechanical activity, akin to the activity of the artisan, in which bodies hit or push against one another, and in general exhibit a tangible set of causal properties and relations that can be pictured concretely in the mind.

After Descartes met up with Beeckman in November of 1618, he evidently became an enthusiast for Beeckman's style of philosophizing; Beeckman's journal notes his new acquaintance as being one of the very few properly to appreciate physico-mathematics and its advantages. Descartes's adoption of physico-mathematical philosophy became his trademark during the following years, along with a determination to systematize his approach to gaining knowledge – knowledge not just of nature, but of everything. Both of these ambitions are represented in his famous publication of 1637, the *Discourse on the Method* together with the three "illustrations" of that method, the "Dioptrics," "Geometry," and

"Meteorology." Descartes had spent most of the 1620s living in Paris and mingling with others of compatible philosophical interests (generally anti-, or at least non-Aristotelian), such as Marin Mersenne, Claude Mydorge, and – when he was in town – Pierre Gassendi. But in 1628 Descartes had decided to seek a degree of seclusion, and went to live in the Netherlands. He had resided there already for several years, therefore, when the *Discourse* and its "Essays" appeared from the Elzevier publishing house in the Dutch university town of Leiden.[5]

The famous "method," which Descartes published here for the first time, represented an attempt to ground all of his ideas in the various sciences on a foundation of certainty. Not for Descartes was there to be a conjectural or hypothetical presentation of the causes at work in the world, as Beeckman's had been; Descartes wanted to present explanations that could not (he hoped) possibly be challenged. In other words, he wanted *certainty* rather than mere *opinion*; his ideas were to be accepted for their *truth*, not simply for their *likelihood* or even mere ingenuity.

Descartes's enormous ambition, in fact, had led him to dream of replacing Aristotle as the master of philosophy. As part of that desire, he sent as a gift one of the first copies of the *Discourse* to his old college at La Flèche, apparently in hope of persuading the intellectually sophisticated Jesuits to use his writings as part of their teaching curriculum. Descartes envisaged the *replacement* of the pre-eminent ancient authority, rather than a humanistic emulation. In the opening decades of the seventeenth century he was by no means alone in this goal; it is of a piece with Galileo's militant anti-Aristotelianism, and strikes a similar note to the work of Pierre Gassendi, another French world-builder of the period. Like Descartes, Gassendi disliked Aristotelian philosophy; unlike Descartes, however, Gassendi still looked towards classical antiquity for his models of philosophical inquiry. Retaining a standard humanist perspective on ancient authority, Gassendi simply substituted the ancient atomism of Epicurus for Aristotle's philosophy, while attempting at the same time to rid it of its atheistical connotations (Gassendi was himself a Catholic priest). Descartes's approach, in rejecting ancient authority completely, thus represents a significant departure from the usual cultural norms.

Seeking the security of absolute certainty for his philosophy was not simply an attractive luxury for Descartes. The attacks on the Aristotelianism of the schools, which had become so frequent by the early seventeenth century, had produced a fearsome armoury of argumentative weapons. The most lethal, particularly in France, where they had proven especially popular, were the weapons of philosophical scepticism. Once again, we have here to do with classical sources: ancient Greek scepticism had already created the chief arguments that were adopted so eagerly in the second half of the sixteenth century and beginning of the seventeenth. Chief among these ancient sources were the writings of the late-antique writer Sextus Empiricus (*c*.200 AD), and his position, known as Pyrrhonism (named after the supposed founder of this philosophical position, Pyrrho of Elis), provided the most destructive weapons of all. Sextus had developed a multitude of standard arguments against the possibility of acquiring certainty in any kind of knowledge whatsoever. These boiled down to two main kinds.

First, knowledge obtained through the senses necessarily lacked absolute certainty because we know that the senses deceive, as any number of optical illusions bear witness. So it is never possible to be completely certain of the truth of anything learned through our senses; we could always be deceived. Notice that Sextus's position was not that we are routinely *likely* to be deceived, in most ordinary situations. Instead, his point was to undermine the pretensions of dogmatic philosophers such as Aristotle, who purported to be able to give absolute demonstrative proofs in philosophy, akin to those of mathematics.

So the second source of knowledge that Sextus attacked as uncertain was human reason, including mathematical deduction itself. In the latter case, Sextus directed his arguments against the formal deductive proofs familiar from Euclid's great work *Elements*. Imagine, says Sextus, checking a deductive proof in geometry by working through every step in the argument so as to show that the conclusion followed infallibly from the starting assumptions. Human beings are not perfect reasoners, he notes, so that in checking a claimed proof, one might mistakenly accept an inferential step as correct when it was in fact false. How could such an error be avoided? One could check the proof several times to make sure of its soundness, of course, but it would still be *possible* to make an oversight on every occasion. One can never, therefore, be *absolutely sure* of the proof's validity. Again, Sextus wanted to show that philosophers who claimed certainty for their assertions were not justified in doing so, however likely those assertions might seem; he advocated suspension of judgement on all issues.

Recall once more that Descartes's ambition was to supplant Aristotle as the pre-eminent philosopher, whose works were studied in all the schools of Europe. Pyrrhonian scepticism was one of the weapons that had been successfully used to weaken the hold of Aristotelian philosophy, precisely by showing that the supposed certainty of its conclusions was nothing but an illusion. Scepticism of this sort was therefore a line of attack against which Descartes wanted his own philosophy to be secure. Given the fundamental nature of the Pyrrhonian arguments, however, this was no easy task. Was there *anything* that could be immune to Pyrrhonian assaults on both the senses and reason? After all, these were not arguments against particular knowledge claims, such as the centrality of the earth in the universe, but attacks on all dogmatic claims to truth whatsoever.

Descartes found himself, then, with a kind of natural philosophy, the physico-mathematical corpuscularism of Isaac Beeckman, that he wanted to argue was superior to any other existing kind. At the same time, he wanted to demonstrate the clear superiority of that natural philosophy over Aristotle's by showing that it was rooted in an absolute certainty that Aristotelianism lacked. His solution, as first put forward in the *Discourse on the Method*, was to convince his reader that the universe is composed of nothing but those things that mathematical magnitudes are suitable for describing, and that causal explanations for all observed phenomena can be provided from *mechanical* principles that fitted such a universe.

On this basis, Descartes hoped that an operationally defined philosophy of nature could be made to appear as a complete natural philosophy, leaving no loose ends.

II Getting inside the mind of God

Without the idea of God, Descartes's remarkable project would have been impossible. In order to be certain of what the world contains, and how it can be spoken of, Descartes needed to circumvent Pyrrhonian scepticism. As we have seen, *refuting* Pyrrhonism was not really possible; its arguments were so fundamental that they automatically applied to any refutations brought against it. So Descartes took a different route towards creating conviction in his reader's mind of the truth of his claims.

His tactic is a famous one: it involved inviting the reader to think along with him so as to become fully persuaded of the truth of his claims. Formal reasoning, one of the targets of the sceptics, was thereby avoided. Rather than wrangling with sceptics, Descartes used their approach as a resource: he begins, in the *Discourse*, by considering the question of how we can be certain of anything at all. He notices how easy it is to find grounds for doubting things, as the Pyrrhonists had long ago found out. Accordingly, he approaches the problem from the other end, so to speak, and asks whether we are in fact left with anything whatsoever if we decide simply to *reject*, as if it were clearly false, anything that could conceivably be doubted – no matter how outlandish the grounds for that doubt might be.

The usual sorts of considerations lead him to reject on this basis all sensory evidence, and even the truths of mathematics itself. Could anything at all be left? Descartes tells the reader of how he became aware of one thing that could not be doubted, even with everything else apparently in ruins. That one thing was his *own existence*: "I think, therefore I am" (*je pense, donc je suis* in this French text, or *cogito ergo sum*, as it appears in 1641 in his Latin work known in English as the *Meditations*).[6] At last, an unquestionable truth – but how to make it do any work? This is the point at which God comes in. In being aware, as his reader should also now be, of his own existence, Descartes is simultaneously aware of his own imperfection. A perfect being, after all, could not be so full of doubts. And the concept of imperfection is itself clearly just the inverse of a concept of perfection; the former concept presupposes the latter. Where, then, did his own concept of perfection come from? Here Descartes says that he could not have acquired the concept of perfection from anything that was itself less than perfect, just as one cannot produce something from nothing; as he argued more formally in the *Meditations*, a cause cannot be less than the effect that it produces.[7] That cause cannot be himself, since he is manifestly less than perfect. So it must come from something outside himself that *is* perfect. In this way, he establishes the necessary existence of a perfect God.

Furthermore, the perfection of God means that He would not mislead us in regard to those things that we perceive "very clearly and very distinctly" (of which the awareness of one's own existence was, of course, the proto-type for Descartes); a propensity to deceive would be an imperfection. Thus, clearly and distinctly perceived ideas must be true. And that was Descartes's refutation of philosophical scepticism.[8]

Physics, Descartes's real quarry, followed hard on the heels of this meta-physical argument. Having established a proper criterion for the truth of ideas, Descartes immediately applied it to matter. Matter and its properties were central to the kind of explanations favoured by Beeckman and now Descartes: above all, matter was assumed to be *inert*. This meant that a piece of matter had no propensity for moving itself – it was, in fact, *dead*. Thus the only way to get it to do anything was to apply to it some outside moving agency.

The existence in a body of sensory qualities such as colour or tem-perature constituted a particularly crucial point of contention between Descartes and the Aristotelians. For the latter, qualities were real things possessed by the objects exhibiting them: a red dress is red because it possesses the quality of redness, much as a rich man is wealthy because he possesses wealth; a fire is hot because of the large amount of heat in it, and so forth. Descartes rejected such a view, holding instead that such qualities are just psychological impressions in the person experiencing them. He had already explained the idea in a little book that he had written, in French, about four years earlier (it was only finally published after his death):[9] in the opening chapter of *Le monde* ("The World"), he had described such phenomena in terms of the purely conventional meaning of *words*:

Words, as you well know, bear no resemblance to the things they signify, and yet they make us think of these things, frequently even without paying attention to the sound of the words or to their syllables. . . . [W]hy could nature not also have established some sign which would make us have the sensation of light, even if the sign contained nothing in itself which is similar to this sensation? Is it not thus that nature has estab-lished laughter and tears, to make us read joy and sadness on the faces of men?[10]

The sounds of words, he argues, cause certain motions in our sensory apparatus and thereby create in our minds the ideas that we have learned to associate with those particular movements. Similarly, he says, our sensation of light could be described as resulting from certain motions created in our eyes, which the mind similarly experiences in a way unrelated to the real nature of the causal agent. Descartes then steps back from these psychologically involved arguments to give another, more direct example of the difference between our sensations and the reality lying behind them. The sense of touch is the most immediate of all, he says, but even here its lessons to us need in no way bear witness to a real quality existing outside us. A feather tickles us; do we then say that the feather possesses within it something that *resembles* that sensation? Descartes makes these arguments in order to draw a formal distinction between our talk of a quality as something that we *experience* and our talk of that quality as a property of the thing experienced. This is because he wants to talk about qualities (specifically, in *Le monde*, light) as something that is really only a property of the motions, or tendencies to motion, of material bodies.

Because Descartes's real goal was to provide a solid philosophical underpinning for his physico-mathematical corpuscularism, he moved in *Le monde* directly from "clear and distinct ideas" and God's guarantee of their truth to the nature of matter, so as to show that matter had just those properties, and *only* those properties, that his favoured kind of physics was capable of discussing. If he could succeed in doing that, then he could claim that his was a comprehensive natural philosophy capable (in principle) of explaining everything. What is matter? The only clear and distinct idea we have of a material body is of its spatial extension: think of a body, and you can imagine it being of a different colour, even of a different shape, of a different temperature, of a different smell, and so on; but your idea of that body cannot dispense with the notion that it is extended in space. Hence, as the only truly clear and distinct idea we have of what a body *is*, and therefore the only *true* idea of the nature of a body, geometrical extension must be what matter really, in itself, is. To use Aristotelian language of a sort that Descartes preferred to avoid, geometrical extension was the *essence* of matter.

In the *Discourse* he only outlined the physics that he had already developed on this basis in *Le monde*. The full publication of the arguments had to await the appearance of the *Meditations* in 1641, dealing chiefly with the metaphysical foundations of his position, and of the *Principles of Philosophy* in 1644, which is a greatly expanded and systematized version of *Le monde*. In pursuing Descartes's world-building further, therefore, it will be necessary to consider the construction of the universe found in those two last-mentioned works.

III Matter in motion

By identifying material substance with geometrical extension, Descartes placed a fundamental constraint on the sort of world that he could build. In effect, he was saying that space and matter are *identical*: where there is one, there is necessarily the other, because they are the *same thing.* Consequently, Descartes's universe could not contain any empty space, because there was no such thing – it was *inconceivable*, and therefore did not exist. From the very beginning, then, Descartes's cosmology possessed properties that had consequences. Descartes built his universe by tracing them.

In both *Le monde* and the *Principles of Philosophy*,[11] Descartes tells a story about the creation and development of an imaginary world. There were theological difficulties with presenting dogmatically an account of the real universe, especially since Descartes wanted to give an account of its gradual formation to yield a world looking just like our own. So he adopts the fiction that his genetic account of the universe is just a fable; an account of how a world just like ours *could* have come into being, even though (he is careful to note) we know that the real universe was in fact created by God just as it appears to us now.[12] This account makes central use of the properties of matter as deduced from his essential definition of it: he starts out with an undifferentiated and limitless expanse of pure extension, which is the same as undifferentiated matter. Left to itself, of course, this would be an uninteresting world in which nothing could ever happen, or individual objects ever come to be. So God is commandeered to introduce motion into this continuum.

The initial disturbance sets everything else in train. Since there is no qualitative difference between any one region of space/matter and any other, Descartes argues that the only kind of differentiation between distinct portions will come about as a result of their motions relative to one another. Furthermore, he is able to say something about the typical *kinds* of motion that will tend to appear in such a circumstance. Because matter is the same as space, it will necessarily be incompressible. This is because if you tried to compress a body, a volume of matter, by squeezing it, then in making it smaller (decreasing the size of a sphere, for example), you would at the same time have left behind a shell of space around it exactly equal to the volume by which the body had shrunk. The shrunken body would consist of less matter (because it would correspond to a smaller region of space), while the matter that it had lost would *still exist*, in the form of the shell of space left behind by its shrinkage.

Because matter cannot be compressed, the movement of any material body will always require that another, adjacent body move out of the way. That adjacent portion of matter will, in turn, have to be made room for by the motion of another equivalent body, and so on. The only way that this could happen, other than having an infinite succession of bodies, each one moving so as to make way for the next, is, says Descartes, if the succession of bodies joins back onto itself in a kind of circle – like the motion of water in a whirlpool.

Vortical motion, therefore, was a fundamental feature of Cartesian physics. It also served to provide immediate empirical plausibility to the world-picture that Descartes was contriving. Tycho had rejected the notion of physically real celestial spheres to carry the planets around in their orbits, as had Kepler. But Tycho had not suggested any replacement for the spheres as the physical explanation of planetary motion, while Kepler had suggested a rather elaborate kind of dynamics to push planets around the sun.[13] Descartes now had a much more intuitively appealing, and obvious, alternative: vortical motion of fluid matter around a central sun would serve to sweep the planets around it much as floating bodies swirl around in a whirlpool.

Before examining some of the details of Descartes's physical explanations, we should first consider his conception of his world system's cognitive status – the character of the knowledge that it was fit to produce. Was it hypothetical knowledge of the real world; knowledge of an imaginary world only; or was it based on grounds so unquestionable that it *must* correspond to the way our world truly is, because our world could *not* be different? Descartes's answer reaffirms his basic commitment to physico-mathematical explanations. The concluding item in Part II of the *Principles* is headed as follows:

> The only principles which I accept, or require, in physics are those of geometry and pure mathematics; these principles explain all natural phenomena, and enable us to provide quite certain demonstrations regarding them.[14]

Descartes goes on to explain what he means by this claim that his explanations are "mathematical"; once again, everything depends on his understanding of the nature of matter.

> I freely acknowledge that I recognize no matter in corporeal things apart from that which the geometers call quantity, and take as the object of their demonstrations, i.e. that to which every kind of division, shape and motion is applicable. Moreover, my consideration of such matter involves absolutely nothing apart from these divisions, shapes and motions; and even with regard to these, I will admit as true only what has been deduced from indubitable common notions[15] so evidently that it is fit to be considered as mathematical demonstration. And since all natural phenomena can be explained in this way, as will become clear in what follows, I do not think that any other principles are either admissible or desirable in physics.[16]

His reasoning, in other words, even about physical things, is "mathematical" in that it partakes of the clarity and soundness of true mathematical demonstrations, and does not even refer to anything that geometers do not include in their own demonstrations. And this is the case because Descartes recognizes as physical phenomena nothing except the behaviours of mathematically defined matter; there is nothing else.

Except, of course, that there is actually rather a lot more. Descartes's account (above, section I) of the relationship between, on the one hand, perceived qualities and, on the other, qualities as properties inhering in physical bodies was a crucial preparatory stage in his argument. What Descartes had to do, in order to make his account of the universe convincing, was to take most of the qualitative characteristics of things, from which our experience of the world is largely created – colours, tastes, smells, sounds, and so on – and displace them from the external physical world itself to our human perceptual apparatus. He could now assert that the correlates of those qualities out in the world have no resemblance whatsoever to our corresponding experiences (unless that resemblance happened to be purely "mathematical," as in the case of a body's shape or size). The door was then open for Descartes to provide such tales as his explanation of colours, which spoke of the relative rates of rotation of the supposed tiny material globules that serve to transmit the pressure that our eyes receive and that our minds experience as sensations of colour.[17] Colour itself only existed in our minds; all else was quantity.

In order to banish qualities from the physical universe, then, Descartes had to consign them to a non-physical realm, that of the human mind. He described the totality of existence as being composed of two kinds of substances: one was matter/extension, which took care of the natural world, and the other was what he called, in Latin, *res cogitans*, "thinking stuff." It was characterized solely by its capacity for thinking, and complemented the physical body of a human being by playing the part of the soul. Descartes stressed that its categorical difference from the stuff of the material world meant that it existed independently of the body; the human soul, that is, did not die with the body but was immortal. Animals, he thought, have no such souls, and can be understood as elaborate automata, like clockwork toys. The human body too could be understood as a machine, even though it housed in addition an immortal, unextended and immaterial soul. Descartes, like many natural philosophers in this period, had a great interest in medicine and the prolongation of life, and attempted quite detailed accounts of how the body is put together and how to understand its operations in mechanical terms.

Connected with his medical interests, Descartes also wrote on a standard theme of the period, the "passions of the soul." His book on the subject, published in 1649, was prompted by his extensive correspondence with an aristocratic admirer, the Princess Elizabeth of Bohemia, for whom Descartes served in these letters as a general medical adviser.[18] Elizabeth was also an acute critic of Descartes's philosophy, and the combination of philosophical discussion and a concern with advising the princess on ways of combating depression led Descartes to develop a quite systematic account of the ways in which the mind was affected by the state of the body (and vice versa) – the "passions" of the soul being precisely these ways in which the soul, or mind, was "passively" affected by external bodily conditions, as opposed to its "active" control of the body by the exercise of the will. Descartes stressed ways of ameliorating the effects of the "passions" by characterizing emotions in physiological terms and considering how to affect those physiological states for the better. And the human body in which these things went on was still, in principle, mechanically intelligible.

Dead matter, activated by motion originally impressed on it by God, exhausted the contents of both Descartes's natural philosophy and the universe to which it referred. But this was possible only by virtue of ascribing all the *other* aspects of the physical world to the inauthentic apprehensions of the *res cogitans*.

IV Believing in Descartes's universe through practical analogies

For Descartes's contemporaries and followers well into the eighteenth century, the importance and attractiveness of his natural philosophy lay not so much in its claims to be rooted in necessarily true assumptions as in the characteristics of its individual explanations. Descartes provided the outline of a very powerful approach to explaining any and all natural phenomena, one that appealed even to people who thought that its concepts were more conjectural than certain.

Descartes's natural philosophy appealed strongly to intuitions derived from the common experience of practical engagement with the world. He uses in his physical writings countless analogies with everyday situations to illustrate, and to render believable, the often highly imaginative mechanisms that he invokes as explanations of apparently very unmechanical phenomena. Light and colour, as discussed in the "Dioptrics," were reduced to pressure in a medium and to rotational tendencies of tiny globules of matter. In fact, in the world-picture presented in *Le monde* in the early 1630s, Descartes made optical phenomena central to his presentation: the full title of the work is *Le monde, ou le traité de la lumiere*, that is, "The World, or Treatise on Light". It is because of his concern with light that Descartes starts out the work with the discussion of the senses considered above. Since he wished to design a world-picture around the behaviour of light, and also to build that world-picture out of explanations that spoke of matter and motion, he needed to establish a bridge between the two. By persuading the reader that light could be fully accounted for by reference to things that themselves had no luminous properties whatsoever, he allowed himself to speak thereafter in terms of structural parallels between the experienced behaviours of light and the independently experienced (and conceptualized) behaviours of material bodies.

Descartes's use of mechanical analogies to make his points is exemplified throughout *Le monde*, but perhaps an example from the slightly later

"Dioptrics" illustrates his style best of all. He there tries to make plausible the idea that vision is accomplished in a manner exactly analogous to the perception of the world achieved by a blind man using a cane – it is essentially a kind of pressure, in this case exerted against the eye, and experienced as light by the unaccountable process of our minds' apprehension of that pressure. Vision is achieved through the action produced by the seen object, an action that is spread from it in all directions. Here Descartes makes use of another analogy, this time with a vat of wine grapes. The weight of the liquid filling the gaps between the grapes exerts a pressure against the walls of the vat that is the result of all the liquid acting together. The contributing action of those parts of the liquid at its surface, for instance, is felt at every point of the vat's walls equally: all the parts of the liquid conspire together to press against every point of the walls. Descartes's non-trivial point here is that the action of light, like the pressure of the wine, tends to operate in straight lines emerging from the luminous body (he cites the sun) in all directions, and that it is a kind of *tendency* to motion rather than motion itself. That tendency is transmitted through a material medium occupying the region between our eyes and the light-source, just as the wine tends to move downwards by an action that is transmitted through the body of the wine. In neither case is there actual motion, but, as the wine example can again show, the phenomenon follows the same rules of behaviour as motion itself. Thus the wine's pressure is manifested by opening a hole at some place near the bottom of the vat; wine will spurt out, regardless of exactly where the hole is made, showing that the *tendency* to motion really does operate towards many places simultaneously. Thus, while bodies cannot *actually* move in different directions simultaneously, they can in this way *tend* to do so.

Descartes's "Dioptrics" is somewhat unusual among his writings, because it uses a variety of such analogies in an openly inconsistent and imprecise way without purporting to present an unequivocal, absolutely true account of how light behaves and why. This is because the text is directed above all towards artisans: specifically, people who possess the skills to grind optical lenses with precision. Descartes wants to enlist their aid in the production of an optical lens that will be free of aberration, by focusing the light passing through it from some point-source all at the same focus point, instead of its being smeared out depending on which part of the lens had refracted it. Throughout his general discussions of the behaviour of light as it is transmitted, reflected, and refracted, Descartes renders his characterizations plausible through the use of various analogies that are meant to convey an idea rather than to prove its validity; by accompanying him, the reader is ultimately led, via a description of the sine law of refraction that introduced it for the first time in print,[19] to a description of a lens-grinding apparatus that is designed to produce a curved surface calculated to produce images free of aberration. It is for the artisan, who might actually build such a device, that Descartes is ostensibly writing,

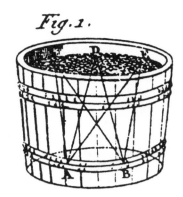

Figure 5.1 *The wine-vat, from Descartes's* Dioptrics.

rather than a philosopher who would want to know whether the things Descartes says are *true* or not.[20] The practical, operational purposes of this optical treatise are therefore paramount; the details of the natural philosophy underpinning them are secondary.

In part this is because the essays following Descartes's *Discourse* were explicitly intended as advertisements for his philosophical abilities, rather than being full statements of his philosophy:

> I thought it convenient for me to choose certain subjects which, without being highly controversial and without obliging me to reveal more of my principles than I wished, would nonetheless show quite clearly what I can, and what I cannot, achieve in the sciences.[21]

However, Descartes used physical analogies almost as much in his formal writings as in these purportedly illustrative accounts of his results. In *Le monde* (which was originally intended for publication), as also in the *Principles of Philosophy* and elsewhere, the very "principles" themselves, not just their applications, are explained and made credible by similar use of everyday analogies.[22]

In *Le monde*, for example, in explaining the possibility of circular motion in a world completely filled with incompressible matter, Descartes invites the reader to consider

> fish swimming in the pond of a fountain: if they do not come too near the surface of the water, they cause no motion in it at all, even though they are passing beneath it with great speed. From this it clearly appears that the water they push before them does not push all the water in the

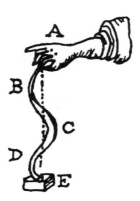

Figure 5.2 *The crooked stick and its straight transmission of force, from Descartes's* Le monde.

pool indiscriminately: it pushes only the water which can best serve to perfect the circle of their movement and to occupy the place which they vacate.[23]

In another place, Descartes uses analogies similar to some of those in the "Dioptrics" to explain his notion of the transmission of light in all directions from the sun. It involves his model of the makeup of the heavenly expanses themselves, which serve as the medium of transmission. Descartes represented the heavens as being composed primarily of roundish globules of solid matter, all in contact with one another like pebbles in a bucket. These globules, which Descartes calls his "second element," communicate the pressure that is the underlying reality of light. Descartes explains how light-rays can appear to travel in straight lines despite the fact that the lumps of second element are not themselves arranged in linear fashion. So, in one case, he uses the example of a curly stick. When the end of the stick pushes against the ground, he says, the pressure is communicated up the stick to the hand at the other end, and the direction of that transmission is a straight line, just as it would be were the stick itself completely straight. So, as the action, or tendency to motion, of light is communicated via the solid globules of second element, it takes place in an overall straight line despite the irregularity of the globules' arrangement.

Descartes dealt similarly even with the most fundamental principles of his physics. As fundamental, they were supposed to be clearly deducible from his already-established metaphysics, but, nonetheless, he still felt

Figure 5.3 *Action transmitted through globules representing particles of second element, from Descartes's* Le monde.

the need to persuade his reader with homely illustrations that would integrate abstract physical principles with everyday experience. Thus, one of Descartes's fundamental "laws of nature" presented in his formal treatise *Principles of Philosophy* held that "all motion is in itself rectilinear; and hence any body moving in a circle always tends to move away from the centre of the circle which it describes."[24] The exposition of this law involves reference to a picture of a hand brandishing a sling that contains a pebble. "For example, when the stone A is rotated in the sling EA and describes the circle ABF; at the instant at which it is at point A, it is inclined to move along the tangent of the circle toward C." This is because "we cannot conceive" that it possesses any of its *circular* movement when it is considered at the single point A; all it then has "in it" is the rectilinear tendency towards C. "Moreover, this is confirmed by experience, because if the stone then leaves the sling, it will continue to move, not toward B, but toward C."[25] (See Figure 5.4.)

Descartes's laws and principles, then, involved frequent and unavoidable recourse to the lessons of everyday experience, rather than being solely grounded in and deduced from formal definitions and formal reasoning. Descartes included in his accounts of the laws governing matter in motion other laws concerning such things as collision between two bodies and their subsequent motions. The metaphysical principle on which he based his laws of collision had to do with God conserving the total amount of motion that He had introduced into the world. Nonetheless, that principle did not enable Descartes to dispense with appeals, whether explicit or implicit, to everyday intuitions about how material bodies behave. Descartes could not have convinced a disembodied pure intelligence of the truth of his laws of nature; he actually needed recourse to what human beings already knew.

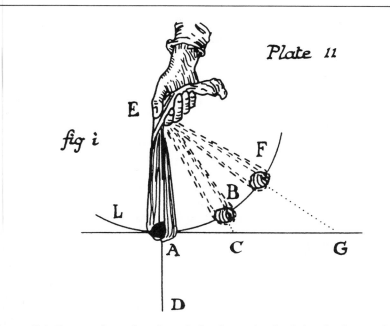

Figure 5.4 *Descartes's explanation of the forces involved in circular motion, from* Principles of Philosophy.

V Descartes's cosmos

The cosmos that Descartes sketched out in his writings, chief among them *Le monde* and *Principles*, was a vast attempt at surpassing Aristotle in comprehensiveness and scope. From his imaginary formation of the universe in *Le monde* and its original injection of motion, he traced the inevitable establishment of huge vortical swirls of matter. He then identified our (Copernican) solar system with one such vortex. The sun is an appearance generated by the presence at the centre of our system of matter that consists of especially small, fluid, and very rapidly moving particles, the incessant jostling of which generates the outward pushing, transmitted through the solid globules of heavenly matter, that we see as light. Indeed, the matter from which the sun is composed is Descartes's "first element," just as the little globules in the expanses of the heavens are his "second element." There is, finally, a "third element," consisting of larger solids, which have no particular characteristic shapes and which compose the earth, planets and comets. Descartes justifies this restriction of the elements to three basic kinds by reference to the theme of light, which structured the account in *Le monde, ou le traité de la lumiere,* and plays the same role regard-

ing the elements in *Principles*. There are three elements because there are
three ways in which matter relates to light phenomena: bodies generate
light, transmit light, and reflect light. Each of these three corresponds to
one of the elements.[26]

The planets, including the earth, swirl around the sun in our solar vortex.
There are also countless other such vortices in the universe: every star that
we can see in the sky is, Descartes maintains, itself a sun at the centre of
its own vortex. The idea of the stars as other suns, with a multiplicity of
worlds dotted throughout a vast, perhaps infinite, expanse of space (which
is matter for Descartes) was by no means new. Apart from classical precedents, more recently there had been the suggestions of a Catholic cardinal,
Nicholas of Cusa, in the fifteenth century, or, in the years immediately preceding 1600, the famous heretic Giordano Bruno, who was burnt at the
stake in Rome for his unorthodox beliefs regarding the Holy Trinity. In
more anti-Catholic times and places than our own, especially in the nine-
teenth century, Bruno was often portrayed, like Galileo, as a victim of
Catholic anti-intellectualism, and the reason for his condemnation implied,
wrongly, to be his unorthodox cosmology.[27]

In any event, Descartes was not worried about the potential heresy inherent in his ideas about the extent of the universe or the nature of the stars.
His major concern, and the one that had persuaded him to suppress *Le
monde* in 1633, was the unorthodoxy (as defined by Galileo's trial) of
holding that the earth is in motion. Descartes published the *Principles*, with
its more elaborate version of the same world-picture as that of *Le monde*,
only once he had thought of a way to deny the movement of the earth
without compromising any of his cosmology. The trick (and that is what it
really was) involved emphasizing the relativity of motion.

In Aristotle's universe, everything had its place. There was a *difference*
between diverse locations that was reflected in the natural motions of
things. The centre of the spherical universe was a unique place with respect
to which motions could be characterized – towards, away from, or around
the centre. It made a difference where something was. Descartes's universe,
by contrast, was designed from the ground up as a mathematical universe,
and as such it mapped directly onto the space defined by Euclidean geometry. Descartes's version of that geometrical space was defined by a great
and lasting mathematical innovation of his own contrivance, which became
known (later) as "analytical geometry." This was first published in another
of the essays, the "Geometry," that accompanied the *Discourse on the Method*
in 1637. The innovative idea was to represent geometrical figures algebrai-
cally: a curve or a solid body could be talked about in terms of the location
of its lines or surfaces relative to three axes at right angles to one another
– axes that Descartes labelled with the letters x, y, and z. A circle of some
radius r, for example, could be represented as a curve in the plane xy
defined by the equation $x^2 + y^2 = r^2$; the circle is here imagined as having
its centre at the origin (where the x and y axes cross).

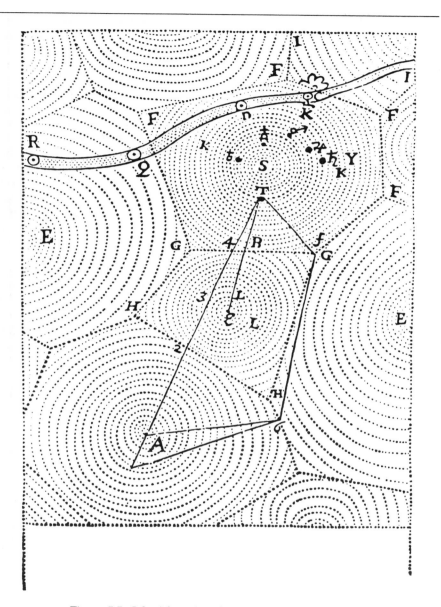

Figure 5.5 *Celestial vortices, from Descartes's* Le monde.

Descartes's conception of the indefinitely extended space framing the cosmos followed exactly the same pattern: it was a space that could always be represented by three orthogonal axes, and the origin of the three axes could be *anywhere you liked*. That is why Descartes's space lacked the absolute character of Aristotle's, where the centre of the universe (and the axes around which the heavens rotated) uniquely defined measurements of position within it. Motion was something real in Descartes's universe, but it was not an absolute, something that could be measured with respect to a unique reference frame. Instead, Descartes defined the motion of a body by reference to the matter through which it was passing. Motion, he wrote in the *Principles*, "is the transference of one piece of matter, or one body, from the vicinity of the other bodies which are in immediate contact with it, and which are regarded as being at rest, to the vicinity of other bodies."[28] Denial of the motion of the earth was then quite easy to achieve, as occurs formally in Part III of the *Principles*:

> since we see that the Earth is not supported by columns or suspended in the air by means of cables but is surrounded on all sides by a very fluid heaven, let us assume that it is at rest and has no innate tendency to motion, since we see no such propensity in it. However, we must not at the same time assume that this prevents it from being carried along by {the current of} that heaven or from following the motion of the heaven without however moving itself: in the same way as a vessel, which is neither driven by the wind or by oars, nor restrained by anchors, remains at rest in the middle of the ocean; although it may perhaps be imperceptibly carried along by {the ebb and flow of} this great mass of water.[29]

The subtlety of Descartes's theology was matched by the subtlety of his physics. As far as he could help it, no one would be able to accuse him of teaching that the earth moves.[30]

VI The success of Descartes's physics

Descartes sometimes described his physics as being, in essence, mechanics, insofar as its explanations were meant to be couched exclusively in terms of matter pushing against other matter much as a weight pushes on the arm of a lever. The success of this "mechanical philosophy," to use Robert Boyle's slightly later expression, was extraordinary, and must not be taken for granted. Why did some natural philosophers prefer Descartes's picture of the world to that of Aristotle?

Any full answer would be enormously complex, but several basic factors can be identified. First of all, Descartes had made it his mission to cover, as far as possible, all the subjects, including specific topics and phenomena, that Aristotle and his subsequent interpreters had themselves discussed. Thus these were the philosophical and natural philosophical questions that people would find familiar from the texts that formed the heart of conventional college and university curricula. Descartes, in effect, wanted to replace Aristotle as the accepted philosophical authority without shaking up the educational structure within which that authority held sway. So, for example, where Aristotle explained the fall of heavy bodies by reference to final causes, relating to the nature of the element earth and its natural place at the centre of the universe, so Descartes too had to explain fall. In his case, the explanation involved the idea that a vortex of (primarily) second element swirls around the earth, and forces ordinary matter, made of third element, in towards the centre of that vortex. This occurs because the *outward* tendency of the revolving second element displaces the more sluggish bits of third element *inwards*. In the case of Descartes's "Meteorology," the third of the essays appearing with the *Discourse*, the very order of treatment of the issues that it discusses follows closely the standard late-sixteenth-century Jesuit commentary on Aristotle's *Meteorology* that was used in Jesuit colleges like La Flèche. Descartes's ambition to replace Aristotle in the schools was in the event largely unsuccessful, certainly in the short run, but his approach meant that people who had been educated in those institutions were perhaps more readily receptive to his ideas.

But alongside the familiarity was also deliberate novelty, and hence unfamiliarity. Descartes had presented a picture of the world that allowed the practice of a kind of physics different from the one pursued by natural philosophers who followed the model laid down by Aristotle. That is, the world-pictures of Descartes and Aristotle corresponded to distinct ways of making explanations. The "mechanical" explanations that Descartes put forward were premised on a foundation of metaphysical certainty for the nature and behaviour of matter. But he himself acknowledged that, so fecund were his explanatory principles, the only problem with explaining specific phenomena was that he could usually imagine several. Determining which among all those possible was the correct explanation was a matter for empirical determination, and even then many explanations would never be able to transcend the hypothetical. The ease of imagining explanations was increased by Descartes's readiness to posit the existence of submicroscopic particles (bits of third element) of whatever particular shape or size he wanted for the purpose. He explained magnets by reference to screw-shaped particles that whirled around between the magnet's poles, completing their circuits by travelling through invisible rifled pores that ran through the body of the magnet – the handedness of the screws determined the difference between north and south magnetic poles.[31] In the "Meteorology," Descartes explains why sea-salt tastes as it does:

> it is not surprising that the particles of salt have a sharp and penetrating taste, which differs a great deal from that of fresh water: for because they

cannot be bent by the fine material that surrounds them, they must always enter rigidly into the pores of the tongue, and thereby penetrate far enough into it to sting; whereas those which compose fresh water, because they are easily bent, merely flow softly over the surface of the tongue, and can hardly be tasted at all.[32]

The atomism of Pierre Gassendi followed a very similar general approach to the explanation of particular phenomena, positing atoms with particular characteristics as and when they were needed. With such a fertile palette from which to work, it is not surprising that this style of explanation became a big hit in many quarters, including those which favoured physico-mathematics.

A generic kind of corpuscularism, making use of *ad hoc* postulated particles invented at the whim of the natural philosopher, appears in many texts dating from the middle of the seventeenth century onwards. One particularly influential example appeared in England, and in English, in 1654; written by Walter Charleton, it bore the unwieldy title *Philosophia Epicuro-Gassendo-Charletoniana*. Its cheerful use of corpuscular accounts of natural (generally terrestrial) phenomena was typical of the pragmatic approach, very different from Descartes's systematicity, that characterized Robert Boyle's approving regard for such ideas starting in the 1650s. Boyle coined the term "mechanical philosophy" to describe all corpuscular-mechanical explanatory approaches, regardless of metaphysical issues such as the difference between Descartes and Gassendi over the existence of a true vacuum (Gassendi allowed one). And Boyle, like Gassendi but unlike Descartes, stressed the *hypothetical* status of these explanations.

One other mechanical natural philosopher of this period deserves mention: the Englishman Thomas Hobbes. As we shall see in Chapter 7, Hobbes's particular conception of natural philosophy bore considerable similarities to the Aristotelian, despite his emphasis on the mechanical intelligibility of physical explanations.

Chapter Six
Extra-Curricular Activities: New Homes for Natural Knowledge

I Changing places

As we saw in Chapter 1, during the earlier part of this period natural philosophy and the mathematical sciences were primarily inhabitants of the universities. One important aspect of the changes that occurred during the sixteenth and seventeenth centuries was the gradual development of significant new locations in which such studies could legitimately take place.

Philosophers of nature were not generally freelancers who conducted their inquiries cut off from the rest of the world. Typically, they possessed identities that were intimately linked to particular kinds of institutional settings. Thus the university natural philosopher had as a primary aspect of his own persona the fact that he was (typically) a professor of some kind, whether specializing in medicine, physics, or (increasingly) astronomy and other mathematical sciences, as in Galileo's case. However, because university natural philosophy and university mathematics usually existed as distinct categories in the teaching curriculum, people whose work and interests tended to blur the distinction between them often had difficulty in pursuing that work within the university's institutional constraints. The clearest examples from the sixteenth century come from astronomy, one of the classical mathematical sciences and an area of inquiry in which physical, cosmological issues frequently became relevant.

As we saw in Chapter 1, section III, during the later Middle Ages the relationship between natural philosophy as applied to the heavens and the mathematical science of astronomy remained rather ambiguous. In principle, the astronomers restricted themselves to calculating the motions of heavenly bodies, an instrumental task that ought not to interfere with the causal, explanatory inquiries of the physicists. But at the same time, physicists could not always ignore the fact that astronomers, in order to perform their own tasks, found particular specifications of planetary orbits and other hypothesized features of celestial behaviour essential to accommodate their data, and that these specifications typically deviated from the simpler models applied in natural philosophy. Thus astronomers all used eccentric circles and epicycles to perform their calculations, and they did not do so as computational shortcuts but as constitutive elements of their work – if they had refused to use them, they simply would have been unable to make quantitative predictions with the same accuracy. So what was the physicist to make of this apparent fact? A popular solution to the problem was, as always, to ignore it. Astronomers took their basic, controlling physical assumptions from the physicists, while the physicists largely ignored conceivable physical implications derived from the work of astronomers.[1]

As we saw in Chapter 2, Copernicus both adhered to this model of how to do astronomy, and deviated from it in spectacular fashion. He modelled *De revolutionibus* on the *Almagest*, even devoting his own Book I, like Ptolemy's, to physical issues concerning the place of the earth in the universe and the question of its motion or stability. Copernicus's most important deviation from precedent, however, lay not so much in his disagreeing with Ptolemy regarding the earth's motion as in the fact that he effectively turned on its head the usual disciplinary relationship between astronomy and natural philosophy. In his preface, Copernicus spoke scornfully of those who professed astronomy ("mathematics") in the schools, whose disagreements and inconsistencies had led him to make his attempted reform; however, his endeavours to lay down new physical constraints for astronomical theorizing – to do, in effect, cosmology – challenged the physicists themselves even while purporting to do better than his hidebound astronomical colleagues. In effect, by trying to present himself as better than his fellow astronomers, he intruded on the professional turf of the natural philosophers and even tried to use astronomical results themselves as arguments for the superiority of the *physical* principles that he put forward.

Thus Martin Luther, the great religious reformer, while no astronomer or physicist himself, condemned Copernicus for his temerity in wishing "to reverse the entire science of astronomy."[2] The Lutheran theologian Andreas Osiander, who wrote an anonymous preface to the *De revolutionibus*, was more explicit, warning of certain scholars who would be "deeply offended and believe that the liberal arts, which were established long ago on a sound basis, should not be thrown into confusion."[3] As the historian Robert Westman has shown, such a remark reflects the fact that Copernicus was not playing the appropriate role of the astronomer, subservient to the natural philosophers, but was trying to set himself in some way over them. This was not how the disciplines were supposed to be organized.[4] And, as Westman has also noted, it becomes especially important, therefore, that Copernicus did not conduct his astronomical work within the setting of a university. Copernicus was not directly affected by the academic disciplinary hierarchy under the thumb of which his university colleagues typically laboured, and was in consequence freer to engage in whatever boundary-breaking innovations he liked. For Copernicus, astronomy, as a deeply humanist enterprise, was a broader field of study than the "mathematics" of the renaissance university arts curriculum. Ptolemy may have in effect subordinated his astronomy in the *Almagest* to the constraints of physical reasoning, but he nonetheless had discussed physics, and to that extent physics clearly, as far as Copernicus was concerned, fell under the purview of the astronomer.

If only from a negative standpoint, then, Copernican astronomy already provides us with evidence of the importance of institutional contexts for the shaping of the intellectual content of scientific enterprises. Copernicus was more readily able to use astronomy to draw physical conclusions about the universe because he was not a university astronomer, obliged to think in terms created by a well-entrenched disciplinary and curricular structure. One can find similar features in the careers of other sixteenth-century astronomers, as Westman has shown: Copernicus's disciple Rheticus is practically alone as a *university* astronomer who took seriously the cosmological claims of the new astronomy (although even he had very pronounced humanistic credentials).[5] Most prominent among the rest, Tycho Brahe, who rejected the motion of the earth but who was fully prepared to challenge established physical assumptions about the heavens, spent his career not in a university but as a recipient of princely largesse: he was supported first by the magnanimity of the King of Denmark, and then, briefly, as Imperial Mathematician to the Holy Roman Emperor. Johannes Kepler, an ardent Copernican, succeeded Tycho in the same court position after Tycho's death. Royal courts were, by the late sixteenth century and early seventeenth century, beginning in many places to provide an alternative venue for astronomers to pursue their work, free from the structural intellectual constraints of university life.

However, the picture is not simply a negative one, whereby places such as courts provided greater freedom than did universities. Whatever the institutional setting, the kind of life that a philosopher of nature could lead within it determined in positive ways too the content of the knowledge that came to be produced. This point becomes increasingly evident as more and more natural knowledge began to appear from the pens of people who were not part of a university.

It should always be remembered, nonetheless, that nearly everyone contributing to this learned culture in the sixteenth and seventeenth centuries, whether working primarily in a university or in some other setting, had been trained to a greater or lesser extent at a university, and was familiar with the kinds of matters routinely taught there. When, therefore, we look at individuals who did not work in a university (a rapidly increasing number in the seventeenth century), we are usually looking at people whose work was shaped and directed by their non-university careers *from the starting point* of their prior academic training. That is why it remains crucial to remember the scholarly standpoints and doctrines examined in Chapter 1: they often explain what the most innovative of ideas, generated in quite different contexts from those of the university, really meant, and the assumptions in relation to which innovations were devised.

II Galileo: from university to court

The career of Galileo Galilei exemplifies these issues with particular clarity. Galileo spent the earlier part of his career, from 1589 to 1610, as a mathematics professor, first for two years at the University of Pisa (in Tuscany), and then, from 1592 until 1610, at the University of Padua, then a part of the Venetian republic. In 1610, however, he resigned from his university post in order to take up a position as Court Philosopher and Mathematician to the Grand Duke of Tuscany, whose court was in Florence. The move was both consequential and symbolic.

At both of the universities in which he taught, Galileo had held the comparatively lowly status of professor of mathematics. Such posts typically paid salaries a lot lower than those received by professors of natural philosophy, reflecting the relative positions of the two subjects in the disciplinary hierarchy. Galileo did not like this. In particular, the kind of mathematics that he practised carried with it (that is, Galileo invested it with) the kinds of presumptions regarding its relevance to understanding the physical world that we have already seen in Chapter 4. That is why Galileo's position when he moved to Florence in 1610 was that of Court *Philosopher* and Mathematician rather than just "Court Mathematician." The latter title, as in the cases of Tycho and Kepler, was a standard one by this time in a number of Italian and German princely courts, and Galileo evidently regarded it as insufficiently august for him. The exact title of his newly-created post had been negotiated between Galileo and the Grand Duke's secretary in 1610, and Galileo was careful to point out in this correspondence that he had "studied a greater number of years in philosophy than months in pure mathematics."[6] That it was important to him to stress his philosophical credentials suggests that, while annoyed at the higher status of his philosophical colleagues at the universities, Galileo had also himself internalized many of the values that justified his inferior position: he too apparently regarded philosophy as more important, as well as more prestigious, than mathematics.

However, note that he had specified, in his rather slighting reference to mathematics in the letter to the Tuscan court secretary, Vinta, "*pure* mathematics." Pure mathematics, as discussed above in Chapter 1, section II, consisted of geometry and general arithmetic, which dealt respectively with continuous and discontinuous quantity. It was a category distinguished from "mixed mathematics," the latter represented in the medieval quadrivium by astronomy and music but in fact including any subject in which quantities *of* something were the proper subject-matter. Thus Galileo left open for himself a personal disciplinary stake in natural philosophy by virtue of his own expertise in mathematics, insofar as he regarded *mixed* mathematics as a legitimate part of a truly philosophical, causal science of nature. Pure mathematics could not count as such, of course, because it did not talk about the changeable natural things that defined physics for Aristotle, but only the unchanging ideal entities of pure quantity.

In moving to Florence, Galileo became more able than before to promote his philosophical agenda. No longer was he subject to the indignity of being subordinated to qualitative physicists, who were not disposed to take mathematical arguments as being central to the resolution of important philosophical questions. Instead, as one who now drew his status not from his circumscribed university position but from his place as a well-favoured client-courtier of the Grand Duke, Galileo could, as historian Mario Biagioli has argued, in effect redraw the disciplinary map that separated mathematics from physics in the academic world.

It is worth pausing to consider exactly what, in concrete *practical* terms, such a shift of social location meant for a scholar in this period. When scholars published their work, their names typically appeared on the title pages of their books together with an indication of their institutional affiliation. For a scholar who worked in a university, this would therefore be a statement of that person's status as professor of philosophy, or theology, or mathematics, at the university in question. A court mathematician, on the other hand, as in the case of Kepler when he published his *Astronomia nova*, specified its author simply as "His Holy Imperial Majesty's Mathematician."[7] Galileo, after his move, was now able to publish using his new title, a title that itself placed him explicitly under the authority and protection of the Grand Duke of Tuscany. This is not to say that the Grand Duke would be taken as actively endorsing anything that Galileo wrote, any more than a university collectively stood behind every pronouncement of one of its professors. It is to say that Galileo, in publishing as the Grand Duke's philosopher, claimed the right to be taken seriously. It was a right that no longer depended on implicitly restrictive university accreditation, but one that derived from political, state power of a different kind. To draw a modern-day analogy, one might say that it was akin to a government scientist claiming the authority of the state for his or her pronouncements – except that in the early seventeenth century, the very institutions of the modern state that provide the modern parallel were themselves only then starting to come into existence.

Galileo's new position, therefore, pushed him in directions different from those that had driven his career during his days at Padua. In fact, as Biagioli shows, Galileo, far from being simply freed of the constraints that had previously plagued him, was now subject to new pressures and new

ASTRONOMIA NOVA
ΑΙΤΙΟΛΟΓΗΤΟΣ,

S E V

PHYSICA COELESTIS,

tradita commentariis

DE MOTIBVS STELLÆ

M A R T I S,

Ex obſervationibus G. V.

TYCHONIS BRAHE:

Juſſu & ſumptibus

RVDOLPHI II.

R O M A N O R V M
IMPERATORIS &c:

Plurium annorum pertinaci ſtudio
elaborata Pragæ ,

A Sᶜ. Cᶜ. M.ⁱᵘ Sᶜ. Mathematico

J O A N N E   K E P L E R O,

Cum ejusdem Cᶜ. M.ⁱᵘ privilegio ſpeciali

ANNO æræ Dionyſianæ cIɔ·Iɔc ix.

Figure 6.1 *Title page of Kepler's* Astronomia nova.

imperatives that helped to determine his subsequent scholarly persona. In particular, the expectations that he had to meet were such as to pressure him into making claims as spectacular and attention-getting as possible.

Galileo had achieved his position in Florence as the direct result of an enormously showy discovery, one that he quickly parlayed into an offering to the Grand Duke that might, he hoped, be rewarded by a court position. The discovery was that of the four chief moons of Jupiter, which he made and then publicized in early 1610. These moons (or, as Kepler soon afterwards dubbed them, "satellites," that is, companions) were the most remarkable results of Galileo's turning the newly invented telescope to the heavens at the turn of 1609/1610. Galileo had heard tell of this new optical device in mid-1609, while visiting Venice, and had hurriedly improvised one of his own. When, later that year, he used a telescope to scrutinize the night sky, he found a number of wonders that formed the subject of his Latin pamphlet of 1610 called *Sidereus nuncius*.[8]

Besides the moons of Jupiter, *Sidereus nuncius* also announced the existence in the sky of countless previously invisible stars. The faint, irregular white band that stretched around the sky and was known as the Milky Way (*via lactea* in Latin; or "galaxy," from the Greek term for milk) turned out, claimed Galileo, to be composed of masses of tiny stars closely packed together: "For the Galaxy is nothing else than a congeries of innumerable stars distributed in clusters. To whatever region of it you direct your spyglass, an immense number of stars immediately offer themselves to view."[9] Of more obvious cosmological significance was Galileo's report on the appearance of the moon. This celestial body, he said, could be seen through the telescope to be rough and mountainous, its irregular surface being more like that of the earth than the smooth, spherical and unchanging orb imagined by those who adhered to the Aristotelian notion of the perfection of the heavens.

But the companions of Jupiter received the greatest space in Galileo's little book, and figured most prominently on the work's title page. Because he had quickly come to regard his discoveries as his ticket out of the university world and into the Tuscan court, Galileo determined that these altogether new planets (as he called them) should be named so as to flatter his intended patron. He therefore initially named them the "Cosmic Stars," to make a pun on the Grand Duke's name, Cosimo II de' Medici. However, Vinta, the Grand Duke's secretary, worried that the specificity of reference to the Grand Duke might easily be missed. Accordingly, Galileo at the last minute changed the name to the "Medicean Stars," thereby making the dedication unmistakeable. The change came so late, indeed, that the printing process was well under way; as a result, first editions of the *Sidereus nuncius* have the description of the moons as the "Medicean Stars" on a strip of paper glued over the previous, rejected name on the title page.

When once established in Florence, Galileo's career was a practically unremitting succession of controversies, major and minor, with other philosophers, almost all of them academic teachers at universities or colleges. As befitted his new celebrity, Galileo continued to present himself and his achievements in a light that would show them off as novel and remarkable. It would not have assisted his value to his patron, the Grand Duke, to have proceeded to write routine texts that did not challenge received views. The *Sidereus nuncius* had immediately made Galileo famous across Europe, and he clearly intended to make good on the promise that had motivated Cosimo to take him under his personal patronage. Patron–client relationships of this kind were not purely one-way affairs.

The ultimate outcome of the story was Galileo's famous 1633 condemnation in Rome following the publication during the previous year of his Italian *Dialogo*, or "Dialogue Concerning the Two Chief World Systems, Copernican and Ptolemaïc". That notorious work provided a wide variety of arguments in support of the earth's motion, all in the guise of a dialogue that purported to represent the matter as a hypothetical question not admitting of certain resolution. The church authorities were not fooled, and Galileo was forced to abjure belief in the motion of the earth and to spend the rest of his life (he died in 1642) under house arrest. His stellar career in the world of noble courts and fashionable readers (the *Dialogo* was written in the Tuscan vernacular of the Florentine cultural avant-garde rather than in the Latin of the schools) had suddenly plummeted to earth. His original patron, Cosimo, was now dead, and Cosimo's successor had much less invested in Galileo's career than had his predecessor. Even Galileo's old friend from Florence, Maffeo Barberini, who in 1623 had become Pope Urban VIII, had more important political worries to consider in the face of conservative clerical opposition than Galileo's philosophical showiness. Galileo's adventures in extra-curricular activity thus came to a sorry conclusion, but it was one that had produced along the way some extremely noteworthy publications. Galileo's great work on mechanics and moving bodies of 1638, the *Discorsi* ("Discourses and Demonstrations Concerning Two New Sciences") was written and published during the period of Galileo's house arrest, bringing to fruition a body of influential work that had begun in the 1590s.[10]

III Patrons and clients

Patronage of mathematicians and natural philosophers, whether by royal princes in the context of an elaborate court or by lesser nobles on a more intimate level, becomes increasingly common during the seventeenth century. Instances of patronage in the later sixteenth century, as with Tycho's position as Imperial Mathematician, or the Duke of Hesse-Kassel's infatuation with alchemy and his financial support of its practitioners, multiply in succeeding decades as more scholars, not just those involved with scientific pursuits, found patrons who used them as private tutors for their families or as adornments of their own status. A number of notable philosophical examples can be found in England in this period, all of them people whose work was of wide influence.

Among the famous English names whose careers need to be understood in the terms of personal patronage are those of Thomas Harriot, William Harvey, and Thomas Hobbes. Harriot and Hobbes both derived their support from their relationships with particular noble families rather than with the royal court, and a kind of retainer-rôle attached to them. Both men acted as tutors to the family offspring, as well as frequently gracing the dinner tables of their patrons. As clients, they dedicated their works to their patrons in the usual manner, and owed at least some of their public presence to the duty of such a client to be a presence in the public intellectual arena. Harriot was associated with the Earl of Northumberland, at the beginning of the seventeenth century, while Hobbes was the (rather aggressive) pet of the Cavendish family during the century's middle decades. In both cases, their work was at least facilitated in a non-university setting by these familiar relationships. In the specific case of Hobbes, his notorious tendency to become embroiled in vicious public disputes over philosophical issues surely owed something, as with Galileo, to the relative institutional independence that resulted from his position. For Harriot, however, there is a noteworthy difference from Galileo that suggests an important contrast between Galileo's involvement in high-stakes court culture and the less intense experiences of lower-level patron–client relationships. Harriot is famous not for his publications (he did not publish), but for the interest of the work that he recorded in private manuscripts or sometimes discussed in correspondence. The most famous example of such work is perhaps his observation of the moon with the aid of a telescope, over a year before Galileo made his own much more famous telescopic observations of the heavens. Where Galileo was to use his discoveries to advance his career through association with the Florentine court, Harriot did nothing of the sort, treating his observations as matters of philosophical interest only.[11]

A more complicated English example, perhaps more similar in some ways to Galileo's career, and one that indicates subsequent developments, is that of William Harvey. His active career spans the first half of the seventeenth century, a period at the start of which he was a medical student at Padua, still the pre-eminent medical university in Europe, and at the end of which he was a man dispossessed of many of his writings and painstakingly collected notes of many years. The loss (in 1642) was due to theft resulting indirectly from his relationship, as personal physician, to King Charles, who was executed in 1649. Harvey is, of course, famous for having invented the doctrine of the circulation of the blood. This was an ambitious and, so to speak, revolutionary claim to make in the early decades of the seventeenth century because of its radical disregard for the medical orthodoxy of the time, an orthodoxy represented by Harvey's own professional community.

Unremarkably for someone destined for prominence in this period, Harvey was well-connected. He finished at Padua, soon afterwards receiving an official M.D. at Cambridge, in 1602; he had been an undergraduate at Cambridge prior to studying at Padua. Harvey then set about applying for admission into the College of Physicians, in London, the membership of which was necessary to practice medicine in that city. He was admitted as a candidate for fellowship in the College in 1604, and in the same year married Elizabeth Browne. This was a canny match: Elizabeth was the daughter of Sir Lancelot Browne, who had been chief physician to Queen Elizabeth up until her death in 1603, and who now served as physician to her successor, James. Sir Lancelot tried to secure a similar court position for his son-in-law, but failed, dying in 1605. Harvey was finally elected a Fellow of the College of Physicians in 1607.

Harvey soon became a prominent member of that body, and was appointed in 1615 its Lumleian lecturer. The position was a significant source of income to him, as well as providing a platform for the dissemination of his ideas in anatomy and physiology. Indeed, it was in these lectures, first given in 1616 and repeated at intervals over the succeeding years, that Harvey first publicized some of his thoughts on the action of the heart. The full exposition of his views, however, appeared in 1628 in a modest Latin text entitled *Exercitatio anatomica de motu cordis et sanguinis in animalibus* ("Anatomical Exercise on the Motion of the Heart and Blood in Animals"). This work argued that, contrary to the teaching of the second-century medical authority Galen, the heart serves to pump blood continuously around the entire body.[12]

De motu cordis displays both its author's professional affiliations, as a Fellow of the College of Physicians, and his ambitions for furthering his position by tapping into the resources exploited by his father-in-law. Thus there are two dedicatory prefaces to the book, one of them addressed to the president of the College, Dr Argent, and the other, appropriately appearing first, addressed to King Charles. The latter adopts the form that was typical of dedicatory prefaces to scholarly works in the sixteenth and seventeenth centuries, one found, for example, in Vesalius's preface to *De fabrica*, directed at the Holy Roman Emperor (and successful in achieving its end of a court position), or Galileo's own preface to the *Sidereus nuncius*, aimed at Cosimo II de' Medici with the rapid success that we have already seen. Thus Harvey praises Charles with elaborate metaphor, comparing the king in his kingdom to the sun in the universe, and both in turn to the heart in the body, the source of heat and life. As Harvey, with perfect self-knowledge, expresses the matter:

> In offering your Majesty – in the fashion of the times – this account of the heart's movement, I have been encouraged by the fact that almost all our concepts of humanity are modelled on our knowledge of man

himself, and several of our concepts of royalty on our knowledge of the heart.[13]

Harvey will therefore take advantage of this situation to press home the fittingness of dedicating to the King a work that will increase the understanding of the heart and, thereby, be of especial "service" to that analogous source of good, the new king of England (who had only ascended the throne in 1625).

In 1629, Harvey was duly rewarded for his loyalty (and for his appropriate family connections) by being made physician to the King and to the royal household. This elevation certainly did not hurt the fortunes of his very controversial physiological ideas. Even after he had ceased to be an official member of the royal household, he still styled himself, in a work published in 1649, "serenissimae Majestatis Regiae Archiatro," that is, "chief physician to his most serene majesty the King."[14] A hint of what this royal patronage did for Harvey is given by a remark that he makes in the same work to vindicate the idea that blood in the veins always travels towards the heart: "In the exposed internal jugular vein of a doe (in the presence of many nobles and the most serene King, my Master), divided in two across its length, scarcely more than a few drops of blood came out from the lower portion, rising up from the clavicle."[15] Harvey here turns the importance of these illustrious witnesses, and even the implied royal provenance of the doe, to powerful account in refuting his critics. This too, in effect, is the purpose of the dedicatory letter to the College of Physicians that prefaced *De motu cordis*.

That dedication makes the point that Harvey would have published the book sooner if it had not been for his fear that he might have been accused of presumptuousness had he not "first put my thesis before you and confirmed it by visual demonstration, replied to your doubts and objections, and received your distinguished President's vote in favour."[16] Just as with his use of "many nobles" to underwrite his assertions about the demonstration using a doe, Harvey in his most famous work uses the authority of the "learned Doctors" of the College of Physicians as a bulwark against critics. Powerful allies *were* powerful to the extent that others would be likely to defer to them; this was one of the (less materialistic) reasons why having an aristocratic patron was so useful to a natural philosopher.

IV Patrons and institutions

Individual clients were not the only beneficiaries of noble patronage. The increasing prominence during the seventeenth century of philosophical groups, whether known as "societies," "academies," or "colleges," was itself typically a product of patronage by an aristocrat. Besides indicating his position as the Grand Duke of Tuscany's Philosopher and Mathematician, Galileo, in his publications that came after the *Sidereus nuncius*, also included among his title-page qualifications the label "Linceo" following his name. This designation indicated his membership in an exclusive group of philosophically minded gentlemen, a group that had originally been formed in 1603 under the patronage of the Marquess of Monticelli (and later a papal prince), Federico Cesi. The group's name was the Accademia dei Lincei, the "Academy of the Lynxes." The name of this natural-philosophical and mathematical society was intended to indicate that its members were acute observers of nature; lynxes were proverbially keen of sight. Galileo became a member in 1611, following a triumphal visit to Rome occasioned by his recent astronomical discoveries, during which the Academy held a banquet in his honour. He seems particularly to have valued this distinction, as his care to indicate it in his publications and in his printed references to other members shows. It was a small and exclusive group, and Cesi's part in it was that of the noble, and legitimating, collective patron; he was what made it respectable.

There were two particularly important functions that the academy fulfilled. One, of especial interest in regard to Galileo, was that it paid for the publication of Galileo's polemical work *Il saggiatore* ("The Assayer") in 1623 and of the great *Dialogo* in 1632, important acts of institutional patronage. The other was its role as the focus for the establishment by Prince Cesi of a major research library. It is important to remember the significance of libraries for natural philosophy even for philosophers who stressed first-hand observation of nature. Francis Bacon in England also emphasized looking for oneself, but that did not prevent him from explaining in some detail the indispensable function of books and writing in building the edifice of natural knowledge. Indeed, Bacon spoke of the importance of written records of experience (his standard term for this idea was *experientia literata*, "literate experience").[17] By this he meant the digesting of observations in an organized system of classification and cross-referencing. Bacon himself, however, made liberal use of books in putting together his own compendia of natural "facts," such as his *Sylva sylvarum* of 1626; this practice is also instanced by his many occasional references to such "facts" in texts like the *New Organon*. For Bacon, stressing first-hand knowledge did not necessarily exclude drawing upon the first-hand knowledge of someone else.

Cesi's Lincean library, indeed, calls to mind Bacon's own desire, during the 1590s, to have the English state establish a number of research foundations, one of which was a library.[18] Bacon's thwarted attempts to establish a system of royal sponsorship for his natural philosophical project represents an unusually clear formal recognition of the potential virtues of governmental patronage. That is really what his fantasy, the *New Atlantis*, is about. The line between the patronage of an individual natural philosopher by a wealthy member of the aristocracy and the support by the state of some kind of organized research endeavour is in this period very evidently only one of degree, not of kind. Besides works concerning the mathematical sciences, Cesi's library also catered to interests in natural historical and medical matters, much along the lines of Bacon's own vision. The library was meant to create the basis for a reformed organization of all knowledge.[19]

The Linceans' publication of works by Galileo formed a part of this project of reform in the knowledge of nature. For our purposes, it is especially important to notice the crucial role of Cesi himself and of the academy that was his instrument of reform. The patronage of Prince Cesi lent an aura of social respectability to the work of the academy, a social respectability that, like his own association with the Tuscan court, meant a lot to Galileo. By his and others' membership in this college of philosophers, the work done under its approving auspices could claim an immediate status as something worth taking seriously. The Accademia dei Lincei stands as a fine example of the way in which the patronage of a powerful individual could also, and perhaps even more effectively, be mediated by an institutional collaborative structure of some kind.

A somewhat later Italian example serves to make a similar point. In 1657, a group of experimenters was formed in Florence under the active guidance of Prince Leopold of Tuscany, brother of Ferdinand, the Grand Duke of the time. This group became known as the Accademia del Cimento, or "Academy of Experiment," and its activities were recorded contemporaneously in a formal journal which was the basis for its sole publication. The book, published in 1667, bore the title *Saggi di naturali esperienze fatte nell' Accademia del Cimento* ("Essays [i.e. trials] of Natural Experiments made in the Accademia del Cimento"). This lengthy title continued appropriately to honour the group's patron: "sotto la prottezione del serenissimo principe Leopoldo di Toscana" ("under the protection of the most serene Prince Leopold of Tuscany"). In fact, the book (intended for private distribution, the gift of a prince, rather than for sale) represents much the most solid reality of this "academy." Indeed, by the time the *Saggi* were published, the members of the group had ceased to meet or to conduct their philosophical activities. The academy never received any formal charter, was never officially disbanded, and merely became a dead-letter when Leopold, made a cardinal, moved from Florence to Rome. There were, with occasional changes of personnel, nine members of the group, calling themselves by the name "Saggiato" (compare with "Linceo"), that is, "trier" or "tester" or, broadly, "experimenter."

The crucial feature of this group as an "Accademia" was its precise relationship with its patron. The fact of its lack of a formal charter, combined with the absence of regular meeting times, shows that it was very much at the whim of Prince Leopold, who called meetings whenever he felt like it. Like his brother, the Grand Duke Ferdinand, Leopold took an active interest in questions of natural philosophy, and participated in the activities of his group of experimenters not just as a distant patron, but also as an experimenter himself. However, features of the *Saggi* indicate the structural conditions of this group and its relations to its patron: the *Saggi* present experimental reports in a style designed to efface the work of particular members in their production. It is, in that sense, a collective publication, recording what the *group* did, regardless of which particular individual did what. At the same time, as already noted, its title page blares out the central role of Leopold himself as the embodiment of the academy's social character. In that sense, the Accademia del Cimento was represented as Leopold, mediated through the work of his clients, the individual members of the academy.

Mario Biagioli points out that one aspect of Leopold's relationship to his experimenters arose precisely from his noble status: experimental work was typically seen as "mechanical" (recall Bacon in Chapter 3, above), and hence risked appearing improper for the participation of a prince. Leopold could, of course, perform experimental work in private, for his own instruction and pleasure, but publication (including explicitly courtly publication, as in the case of the privately distributed *Saggi*) had to take account of his public image and status. By representing the various philosophers with whom he worked and corresponded as constituting an "academy" that operated under his patronage, Leopold in effect insulated himself from the messy work itself, while still receiving the reflected glory of its achievements. The work of the group, among whose most important members were Giovanni Borelli and Francesco Redi, included a multiplicity of experimental trials of phenomena concerning such things as hydrostatics, barometrics, or the nature and behaviour of heat, all with often expensive apparatus, such as tailor-made glassware. Leopold ensured that the philosophical claims made on the basis of this work were as undogmatic, and as focused on phenomena themselves (rather than causal explanations for those phenomena) as possible – hence reducing the risks attendant upon controversy.

The two most important organized groups of natural philosophers in the seventeenth century, both established in the 1660s and surviving, with modifications, down to the present day, were the Académie Royale des Sciences (Royal Academy of Sciences), in Paris, and the Royal Society of London. Both groups, as their names indicate, were formed under royal patronage, but in their differences further illustrate the fuzziness of the line separating individual patronage from collective, institutionally mediated patronage of natural philosophy in this period. The creation of new venues for the study of nature was a process that required careful adaptation to existing norms of social status and respectability in order to be successful as a rival to the universities.

The Academy of Sciences was founded in Paris at the end of 1666. It was, in effect, invented by Colbert, the chief minister to the King of France, Louis XIV. The basic political strategy of Louis and his ministers was to establish the monarchy as an unrivalled locus of power in the state; that is, their ambition was to set up an "absolutist" state whereby everything was, at least in principle, dependent on central state control. One might analogize it to "totalitarian" regimes in the twentieth century, where much the same ambition was pursued, although generally with greater success. Louis XIV was called "the Sun King" precisely in order to emphasize the idea that he was the source of everything that happened in his kingdom. William Harvey's dedication of *De motu cordis* to England's King Charles had promoted the same image in regard to Charles, whose own attempts at absolutism were finally thwarted by parliament: Harvey had written that the king is "the basis of his kingdoms, the sun of his microcosm, the heart of the state; from him all power arises and all grace stems."[20] Likewise, Colbert was concerned, in the 1660s, to ensure that such an image would apply to Louis. This meant tying all potentially independent sources of power and glory to dependency on the king. In the specific case of the sciences, natural philosophical and mathematical, Colbert wanted to group together the leading practitioners into a formal academy that was institutionally and officially an arm of the state. It was this plan that quickly eventuated in the formation of the Royal Academy of Sciences, inaugurated in December of 1666.

As a serious arm of the state, the Academy, unlike the Accademia del Cimento, possessed a formal constitution that included salaried positions for its regular members and expected duties for them. Those members were drawn not only from around France, but also from abroad; the star of the early Academy was the Dutchman Christiaan Huygens, whose renown as a mathematician and admirer of Cartesian philosophy had already raised him to the premier rank of European physico-mathematicians. Also among the members was a prominent Italian astronomer, Giovanni Domenico Cassini, the first of a long line of Cassinis who dominated Parisian astronomy down through the eighteenth century. In total, there were fifteen original members of the Academy, twelve of whom were French (the third exception was the Dane Ole Rømer). Promise of generous salaries acted as a powerful draw to Paris and to the Academy for these people; Huygens, for example, not only received his official salary, but was also given apartments in the Louvre palace. No expense was spared for the King's (and thus the state's) philosophers of nature.

There were also strict rules regarding the scheduling and conduct of the meetings. When in regular session, the Academy met twice a week, on Wednesdays and Saturdays. These two time-slots were designated for each of the two sections into which the Academy was divided; a division that speaks volumes regarding the basic conceptions of natural knowledge that were taken for granted in the Academy's creation. One section was "mathematical," the other "physical," corresponding surprisingly closely to the disciplinary division that had long been institutionalized in the universities, and that reflected in turn an Aristotelian view of the difference between those two realms of natural knowledge. Thus the "mathematical" section was devoted to the mathematical sciences as we have met them repeatedly before: not just (or even especially) pure mathematics, but also, and indeed more centrally, "mixed" mathematics; that is, all those areas of study that had, during the course of the seventeenth century, become labelled by many practitioners "physico-mathematics." Thus mechanics, astronomy, optics, and all the usual classical mathematical sciences of nature (most practised, to some degree or another, by Huygens) fell under the auspices of this specialized section, leaving to the "physical" section everything else to do with the study of nature – that is, as in the Aristotelian model, all qualitative studies of the natural world, from natural history and chemistry to anatomy.

Although Huygens was fond of postulating mechanical causal explanations for physical phenomena in a broad, non-dogmatic mechanistic idiom, he nonetheless was located in the mathematical rather than the physical section. However, such apparent blurring of boundaries does not seem particularly to have threatened this fundamental division of the early Academy. Huygens's work, after all, was performed in primarily physico-mathematical areas where mathematical relationships between quantities were empirically determinable; this work also, typically, promoted causal explanations for physical phenomena.[21] It was, as we saw in Chapter 4 above, precisely this claim on physical explanation that had motivated the widespread adoption of the new category of physico-mathematics, which in the Academy retained its specifically mathematical mantle. Huygens had written, in 1666, that "the principal occupation of the Assembly and the most useful must be, in my opinion, to work in natural history somewhat in the manner suggested by [Lord] Verulam [i.e. Francis Bacon]."[22] After all, natural history in the Baconian sense was the general gathering of facts about nature, and was thus itself exempt from further meaningful disciplinary subdivisions.

The sectional division between "mathematical" and "physical" was in any case not intended to set up an unbridgeable barrier between two distinct groupings within the Academy. The members of one section were expected to attend the meetings of the other section as well as those of their own. Another example of the planned integration of the Academy's efforts was the most elaborate early state benefaction to the Academy, the Observatoire de Paris (Paris Observatory), opened in 1672 and still standing today. This was an astronomical observatory intended for the use of all the Academy's *savants* (i.e. "learned men"), its basements being suited to laboratory work. In practice, however, these savants tended to be restricted to Cassini and his astronomical assistants: it was the Academy's laboratory in the Bibliothèque du Roi that witnessed anatomical, botanical, and other kinds of natural-philosophical research in the late seventeenth century rather than the Observatory, which remained, therefore, strictly "mathematical."

The intended collectivity of the Academy was also expressed in its intended practice of publishing its work, much like the Accademia del Cimento, under the auspices of the Academy itself rather than under the names of the specific author or authors who had performed the work or written the text. The practice was never entirely satisfactory, however (especially to the specific authors), and was gradually abandoned. This was done in the first instance by exempting certain mathematical and hypothetical treatises from the rule, and substituting instead a process of peer review that enabled the Academy's approval to be granted without removing the writer's own name. This practice also served to distance the Academy collectively from an appearance of endorsing the necessary truth of the hypothesis involved.

It is clear that the relationship between the Academy of Sciences and its ostensible royal patron differed considerably from that between the Accademia del Cimento and its patron. Where Prince Leopold was himself directly involved in the experimental work of his unchartered academy, and, from his perspective at least, the group acted as a mediation between his interests and the world at large, King Louis XIV had very little apparent interest in the work of his own academy. In a sense, it was another government department, with official functions (such as, from about 1685 onwards, assessing patent applications for inventions, a duty formalized in new regulations introduced in 1699). Louis XIV never made a visit to his Academy until fifteen years after its inauguration, despite a fictitious engraving of the new Observatory that symbolically suggested otherwise. The Academy was for Louis just another representation of his power and glory: "l'état, c'est moi," he famously asserted, "the state – that's me!"; the Academy of Sciences was itself just a part of the state, and its achievements just a part of the king's. In a sense, the Academy of Sciences instantiates at the end of the seventeenth century the vision of state science promoted by Bacon at the beginning.

Ironically, the most vehemently Baconian of the seventeenth-century scientific societies instantiated this vision of state-science much less fully than did the Academy. The Royal Society of London for the Improving of Natural Knowledge was founded in the early 1660s, informally in 1660, and receiving royal charters in 1662 and 1663. The latter charter is, in fact, the one under which it still operates today. Its distance from the more centrally initiated bodies that we have considered may be seen in the manner of its formation: whereas the Accademia del Cimento was a result of Prince Leopold's own active interest in the work of various philosophical clients and hangers-on of the Tuscan court, and the Royal Academy of Sciences was established as an official state cultural organization, the Royal Society was constituted by its founding members as a society of like-minded individuals who wanted to conduct on an organized basis experimental and natural historical inquiry, with a rhetorical stress, at least, on potential utility. Only after they had come together as a group did they succeed in acquiring for themselves the royal approval that enabled them to become the Royal Society, and even then they received very little beyond the title itself: no state support of a material nature was forthcoming.[23]

The immediate forebears of the Royal Society were a group meeting in London in 1645, and a group organized in 1651 at Oxford, calling itself the Experimental Philosophy Club, which had a degree of common membership with the earlier London group. The year of the Royal Society's organization, 1660, was also that of the Restoration of the Monarchy in England, following the Civil Wars of the 1640s and the interregnum rule of Oliver Cromwell in the 1650s. One of the early Society's leading figures, John Wilkins, had been brother-in-law to the deceased Cromwell, although that association did not prevent him from being made a bishop in the Church of England in 1668. Wilkins, who had been installed as Warden of Wadham College, Oxford by the victorious parliamentarian regime, had been a member of the Oxford Experimental Philosophy Club in the 1650s, one of a number of leading lights in the later Royal Society who had lived and worked in Oxford during that period. A striking feature of the early Royal Society was its carefully non-partisan character: former royalists rubbed shoulders with parliamentarians, and even Anglicans with Catholics; an unusual kind of ecumenism that emphasized the Society's collective determination to turn away from divisive issues of politics and religion. As Thomas Sprat, in the official *History of the Royal Society* of 1667, put it, the Society's determination was to put aside "the passions, and madness of that dismal Age," referring to the period that had preceded the Restoration of Charles II.[24]

So far was the Royal Society from being a personal patronage interest of the monarch that the King notoriously referred to its members as "my fools," and allegedly mocked them for attempting to weigh the air. In the political atmosphere of England, in which attempted royal absolutism had failed, a broad distribution of authority and of autonomous domains of activity characterized the relationship between the state and the royally sponsored scientific society – just as the opposite was the case in absolutist France.

Conformity with the new political settlement of the Restoration was the watchword of the Royal Society; it had its royal charter, to emphasize its political orthodoxy, while at the same time practising a kind of tolerance meant to avoid becoming too much identified with any particular polarizing political position. Of course, despite this, it was anything but open to talents from anywhere: in many respects it functioned as a gentleman's club, restricted in that way to men, and to men of a certain social class. Both points are nicely illustrated by a famous event of the Society's early years: a visit by Margaret Cavendish, the Duchess of Newcastle. Not only was she a member of a family with significant ties to English natural philosophy in the seventeenth century, Margaret Cavendish was herself the author of treatises on natural philosophy; albeit with an anti-experimental slant analogous to that of Thomas Hobbes, which we shall see in Chapter 7, rather than being consonant with the experimental focus of the Royal Society. Nonetheless, and despite being a woman, she was allowed to attend a meeting of the Society, and, indeed fêted, in 1667. She qualified for such treatment by virtue of her aristocratic status, which made her the social superior of most of the men belonging to the Society. Similar considerations enabled Queen Christina of Sweden, or the Princess Elizabeth of Bohemia, to command the attention of Descartes (in the case of Elizabeth, to Descartes's considerable intellectual benefit). But such deference was always severely limited: an English contemporary, Lady Anne Conway, was a philosophical writer of some note, but like the Duchess of Newcastle was never able (unlike Conway's brothers) to attend one of the universities.[25] Notwithstanding her aristocratic social status, as a woman Margaret Cavendish was never so much as considered as a potential Fellow of the Royal Society, despite her publications on natural philosophy. The Royal Society's institutional form in fact effectively followed the masculine corporate models of Oxford and Cambridge colleges, or the College of Physicians, or the established Anglican Church, or parliament; there was nothing socially revolutionary about the Royal Society.

Perhaps ironically, the Royal Society attempted to legitimize its role in English society by actively exhibiting the diversity of its membership (despite the considerable limitations just mentioned). The alleged lack of partisan interest – of whatever kind – served to present the Society as politically safe, both from the perspective of the government and also, at least equally importantly, from that of already established groups, jealous of their own corporate rights. Such groups included the universities (Oxford and Cambridge, still the only two universities in England), and the College of Physicians, quarters from which it nonetheless received criticisms in its early years. The Society's major platform for propaganda was Thomas Sprat's *History of the Royal Society* (1667), which also placed emphasis on another claimed feature of the Society, the potential usefulness of the knowledge that it was dedicated to creating. Central to its self-image, and of the image that it projected to outsiders, was the Society's adherence to the project of Francis Bacon, Lord Verulam.

Bacon's stress on the utility that his kind of natural philosophy would surely bring to all mankind, and in particular to the English nation, was something that the Royal Society took to itself as it ensured that Bacon's name was continually associated with its own. Baconian rhetoric had been culturally very successful during the 1640s and 1650s, becoming associated particularly with reform projects promoted during the years of the Interregnum by people with anything but royalist sympathies. Nonetheless, Bacon's name remained one to conjure with even after the Restoration, and his close association during his career with monarchical régimes no doubt assisted in making him a safe emblem in the 1660s. The frontispiece to Sprat's *History* itself stands as evidence of this fact: two seated figures are displayed prominently, one on each side of a bust on a pedestal. The bust is of Charles II, the royal patron of the Society; the pedestal acclaims him as the Society's "Author & Patronus" ("Author and Patron," a characterization that attributes the spark and motor of the Society to the king himself, in a rather dim reflection of the "sun king" motif). One of the two seated figures is the Society's president, Lord Brouncker, a politically appropriate individual who had been in exile with Charles in the Netherlands in the years before the Restoration. The other figure is Francis Bacon, Lord Verulam, labelled as "Artium Instaurator" ("Renewer of the Arts," evidently referring to both the liberal and the mechanical arts – pretty much all of knowledge, in fact).

Sprat's book attempts to vindicate this grandiose image by citing some of the things that Fellows of the Society had already accomplished during its few years of existence. The major public face of the Royal Society, however, was a journal. The *Philosophical Transactions* started life as a private, money-making venture by the Royal Society's secretary (until his death in 1677), Henry Oldenburg. Oldenburg was an expatriate German who had lived in England since 1653 and conducted a large philosophical correspondence. The *Philosophical Transactions,* begun in 1665, grew out of his work as an "intelligencer" (to use the seventeenth-century term), but it was immediately seen as the Royal Society's own journal. This was due to the fact that Oldenburg's correspondence was now that of the Society's secretary, and the letters of which he made ample use in the journal were ones that he now received in that official capacity. Nonetheless, the *Philosophical Transactions* did not itself become the official journal of the Royal Society until the middle of the eighteenth century, since when it has been called *Philosophical Transactions of the Royal Society.*

Oldenburg's prefaces to the annual volumes of the journal use Baconian rhetoric lavishly, stressing the utility of the work published in the *Philosophical Transactions'* pages and the way in which that work was focused on the gathering of empirical facts of the kind championed for natural philosophy by Bacon. One symptom of this would-be Baconianism is the scarcity of articles in the journal that might have appeared too theoretical, hypothetical, or speculative. (Chapter 7 will examine a particularly notable instance of the trouble that this policy could cause, one involving Isaac Newton.) However, the same can to some extent be said of the Royal Society itself; Oldenburg's editorial predilections were in step with those of other leading Fellows. The minutes of the Society's meetings display very much the same interests as those represented in the *Philosophical Transactions,* and the meetings regularly involved the reading and discussion of letters to the Society that Oldenburg subsequently printed in the journal.

The literary output of the Society should also be taken to include books published, with the Royal Society's official *imprimatur,* by its Fellows. Those by Robert Boyle were probably the most widely read throughout Europe,

Figure 6.2 *Frontispiece to Thomas Sprat's* History of the Royal Society.

and the ones most often taken by contemporaries to be representative of the Society's own project. (It is also worth noticing that Henry Oldenburg himself did a good deal of work as, in effect, Boyle's publisher during the 1660s and 1670s). Most such books were not publications of the Society itself, but among the small number of exceptions two are especially noteworthy: Robert Hooke's *Micrographia* of 1665, and Isaac Newton's great *Principia* of 1687. *Micrographia* was the first illustrated book of observations made with a microscope (the book *Experimental Philosophy*, by Henry Power, another Fellow of the Royal Society, appeared the previous year, but is mostly bereft of illustrations). The painstaking engravings, some of which may have been made by Christopher Wren, another Fellow as well as architect of St Paul's Cathedral in London, display the strangeness of the world as seen under the microscope, as Hooke details the surprising appearances of products of art (fabric, the points of needles, printed letters in books) and of nature (the "cells," as he dubbed them, visible in pieces of cork, or the detailed anatomy of tiny insects). Art was surprising for the discovered coarseness of objects usually regarded as fine and delicate, whereas nature was remarkable for its hidden delicacy. One was the handiwork of men, the other of God; therein lay the difference.

The publication of Newton's *Principia* was, in fact, beyond the means of the frequently impecunious Royal Society. One of the great gripes of some of the Fellows was that they, unlike their counterparts in the French Academy of Sciences, received no state subsidy, and relied on members' dues. Such reliance was dangerous, because these monies were frequently in arrears, when paid at all. A cash shortfall in the mid-1680s obliged Edmund Halley (after whom the comet was later named) to pay for the publication of the *Principia* out of his own pocket; he was reimbursed with copies of the Society's recent budget-busting publication of Francis Willoughby's *Historia piscium* ("History of Fishes", "history" here meaning a compendious "natural history" in Bacon's sense).

The publishing function of a society of the kind that this section has been considering was perhaps one of the most characteristic features of the new social forums for the sciences in this period. The independence of such societies from the universities (if not from their patrons) was reinforced every time a book appeared with the stamp of their approval; furthermore, when, as was often the case, their patronage devolved from the state (in the person of a princely patron), those books carried the weight and authority not simply of intellectual orthodoxy, but of the political orthodoxy of a temporal power.

Finally, one other "place" in which nature could be studied deserves mention: the home. Scholars have recently begun to look at the ways in which domestic space was utilized for such purposes in the early-modern period, including the roles of women and servants (the most marginalized human components of formal knowledge-making in this period). Easy to overlook, the home and its functional divisions created important focuses

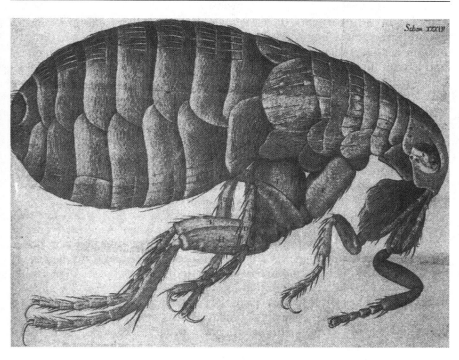

Figure 6.3 *A fold-out view of a flea, from Hooke's* Micrographia.

for knowledge-making that intersected in obvious ways with the social status of the natural philosopher, and hence with the resources (material and social) available to render the knowledge so produced credible to others.

V Institutions for conquering space: natural history and the European embrace of the globe

The Spanish explorations of the Americas, together with voyages from various countries to other unfamiliar regions both east and west, had been an expansion of the European world not only geographically but also intellectually and logistically. In the sixteenth century, a map of the world came to contain more accredited *places*, real places to which travel was now possible; but it also designated previously unknown *things* whose very distance created special problems of accommodation. How could a European know this transformed world? Hitherto, the writings of the ancients had delineated almost everything that was believed to exist: the astronomer Ptolemy, in his great work the *Geography*, had provided the foundations of geographical knowledge, but his world was largely restricted to Europe and the Mediterranean lands. The recent voyages had revealed much more, and expanding world-wide trade required that it be known and controlled. The writings of such authorities as Aristotle and his successor Theophrastus, which had played similar roles in understanding the fauna and flora of the world, had similarly become radically incomplete. One problem that this situation raised was whether the old framework represented by these classical texts was still appropriate; whether it sufficed simply to add new kinds of plants and animals to the old interpretive schemes. Another problem concerned the making of the new knowledge: in natural history, on whose authority and by what mechanisms were descriptions of new organisms to be received into the body of accepted learning?

A European science that would encompass the world needed to bring that world home. If knowledge of such places as South America were not transported back to Europe, that knowledge could not enter the storehouse of generally available truths that constituted the expanding legacy of European learning. In the main, it was knowledge of local particulars rather than of universal truths that highlighted this problem – knowledge of particular species of plants or animals, or even perhaps rock-types, that were only found across the seas. During the later sixteenth century and early seventeenth century, the Jesuit order constituted perhaps the most elaborate and well-organized international network in the world. Jesuits on missions to far-off lands such as China or Canada – unusually, not overtly intended to facilitate trading relationships – were required to report back regularly to their European masters, and the existence of this resource was exploited in some quarters to acquire knowledge of nature for philosophical purposes. In the seventeenth century, the most important figure to make use of the Jesuit network in this way was Athanasius Kircher (1601–80), a member of the Jesuits' Roman College (*Collegio Romano*). Kircher's interests were boundless, and his enormous correspondence with Jesuits around the globe involved, besides matters of natural historical and scientific concern, discussion of languages and cultural practices found worldwide and at different times in the past. Kircher's role as the principal node in a network of (largely Jesuit) correspondence, together with his many publications, rendered him of the first importance in facilitating the organization of apparently miscellaneous reports from abroad. He produced on the basis of such material, for example, the first map depicting the world's ocean currents. Kircher's endeavours were too dependent on his own individual efforts to establish any long-term institutional continuity in themselves, and much the same is true of later Jesuits who took advantage of the same features of their Order's organization for scientific purposes. That such work was possible, however, illustrates the close relationship between administrative networks and scientific networks, which possessed their own capacities for generating global, integrated knowledge. The Jesuit missionary Matteo Ricci in the early seventeenth century used his European mathematical expertise (drawing on the Jesuit tradition in mathematical sciences that we saw in Chapter 4) as a means of ingratiating himself with the Chinese court, using science as a means of furthering diplomatic as well as spiritual endeavours. Trading and associated military networks established by governments or by governmentally sanctioned organizations such as the Dutch East India Company (founded in 1602) are further examples of essentially the same phenomenon.

The trading impulse thus appears as a major force behind the institutionalization of a kind of knowledge that aimed at extending the grasp of Europe over the entire world. The Dutch, who became a major mercantile force starting in the late sixteenth century, and the French, provide clear examples of the role of geographical expansion for the more general expansion of the management of technical and natural historical information in this period. Before examining those cases, however, a glance at the situation that had grown up in parts of Italy during the sixteenth century will show some of the options newly available to other European powers.

Museums of natural history, as well as botanical gardens, appeared in profusion in Italy from the middle of the sixteenth century onwards. The meaning of these collections, however, is not as simple to determine as might at first sight appear. Ulisse Aldrovandi built up a private museum collection in the second half of the century as well as initiating a botanical garden at Bologna. The latter enterprise is easier to explain than the former. Botanical gardens had a long history, having formerly been associated, for example, with monasteries. Their *raison d'être* was pharmaceutical; herbal remedies were staples of contemporary *materia medica*. The first of the new Italian botanical gardens was established at the University of Pisa in 1543, and during the next couple of decades additional examples, including Aldrovandi's, sprang up at several other leading Italian universities. The establishment of new botanical gardens in the sixteenth century was in part a response to the sudden availability of increasing numbers of new plant species that were arriving in Europe, especially from the New World of the Americas. But this process at the same time risked overstepping the boundaries of ancient authority, by introducing plants that had of necessity played no part in the authoritative tradition of medieval medicine. Aldrovandi's natural history museum itself illustrated the general problem: natural history as an intellectual enterprise was rooted in ancient texts, but its bounds were being expanded by the voyages of discovery and their aftermath. What guide for collectors like Aldrovandi could the ancient texts still provide? And what material form would the museums take?

It is somewhat ironic (although scarcely surprising) that the general response by museum-makers to the new worlds discovered to Europe in the sixteenth century was one of assimilation to existing cognitive models. Master narratives of world history and human history, rooted in the Bible, continued to be used in attempts to fit newly-discovered peoples into a pre-existing understanding of human origins and dispersal on the earth. Were the peoples of the Americas related to the Lost Tribes of Israel? How were they related to Noah's sons, who had repopulated the world after the great Flood? Or were they (a rare, because heretical, option) descendants of people who had lived before Adam, separated from the main line of biblical genealogy? In much the same way, natural historians did not usually regard their new subjects as violating classical norms: new plants and animals were still dealt with using the ancient models. Of these, the most congenial was that provided by the first-century Roman writer Pliny the Elder in his work *Natural History*. This text had the advantage of failing to be very systematic at all, so that it set up few generalizations that new discoveries might violate. Instead, it provided a model for how to talk about new species; how to describe them, including discussion of such matters as their cultural or emblematic meanings and their practical uses.

The chief French institution devoted to natural history was the *Jardin du Roi* ("King's Garden," also called the *Jardin des Plantes*) in Paris. It was founded during the first half of the seventeenth century, and followed the pattern of the Italian botanical gardens – that is, it was originally designed for medical purposes. The first French emulation of the Italian botanical gardens, with their public university affiliations, had been the establishment of a garden at Montpellier in 1593 by the French king at the time, Henri IV. Its Parisian counterpart was created, against constant criticism from the Faculty of Medicine, through the endeavours of Guy de La Brosse, a physician to Louis XIII. (The main royal edict for its foundation was issued in 1635.) Like its Italian forebears, the *Jardin du Roi* was intended to preserve and propagate a wide variety of plants for their medicinal uses, and from a scholarly perspective its mission was also to take account of the large number of medicinal herbs that had not been discussed by the ancient Greek botanist Dioscorides – either because they were restricted to northern Europe, or because they came from the New World. As with the example of Pliny, however, that of Dioscorides was not discredited by the novelties: what might appear as shortcomings were usually treated simply as calls for further emulation. Since Guy de La Brosse was also a champion of the new, and unorthodox, chemical remedies that had been promoted during the sixteenth century by Paracelsus, the *Jardin* was also equipped with a chemical laboratory.

The *Jardin du Roi*'s first published catalogue, in 1636, records a stock of more than 1800 different plants. This large number indicates the practical difficulties of taxonomic management that were beginning to engulf European botany, and that would lead by the end of the seventeenth century to

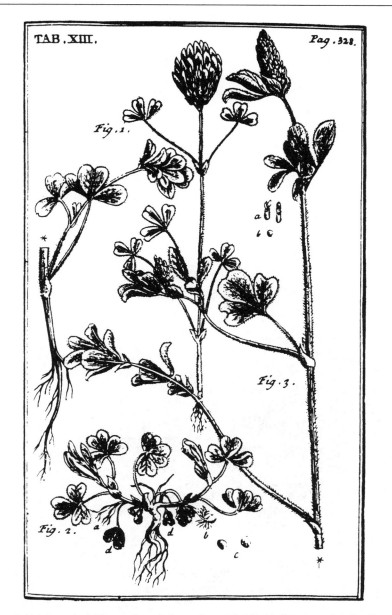

Figure 6.4 *A botanical illustration of clover, from John Ray's* Synopsis Methodica Stirpium Britannicarum *(1st edn, 1690; illus. in 3rd edn, 1724), showing both naturalistic representation and taxonomically significant characteristics.*

128

Perfectus, qui petalis, stylo & staminibus constat; estque vel

 Simplex, qui in flosculos non dividitur, isque vel

  Monopetalos, qui unico petalo sive lamina continuâ constat, ut in *Convolvulo*, *Campanula*, &c. éstque vel

   Uniformis, qui dextram partem sinistræ, & anteriorem posteriori similem, inferiorem superiori dissimilem obtinet, ut in *Convolvulo*. Estque margine vel

    { *Integro*, ut in *Convolvulo*.
    { *In lacinias fisso*, differentes

     { *Numero*, in nonnullis sci. tres, in aliis quatuor, vel quinque, vel sex laciniæ sunt.
     { *Figurâ*, vel angulosâ, vel rotundâ.

   Difformis, cujus non tantùm superiora ab inferioribus, sed & anteriora à posticis differunt, éstque vel

    Semifistularis, ut in *Aristolochia*.

    Labiatus, labio
     { *Unico*, eóque vel *superiore*, ut in *Acantho sativo* ; vel *inferiore*, ut in *Scordio*, &c.
     { *Duobus*, superiore vel

      { *Reflexo* sursum, ut in *Chamæcisso*.
      { *Convexo* sive deorsum reflexo, sive galeato, ut in *Lamio* & plerisque Verticillatis.

    Corniculatus corniculo seu calcaneo concavo & impervio retrorsum extenso, ut in *Delphinio*, *Linaria*, &c.

  Polypetalos sive multifolius est qui pluribus petalis in unica serie aut circulo dispositis componitur ; éstque vel

   Uniformis, in quo petala, figurâ & situ conveniunt, quamvis magnitudine interdum differant ; éstque vel

    { *Dipetalos*, ut in *Circæa Lutetiana*.
    { *Tripetalos*, ut in *Plantagine aquatica*.
    { *Tetrapetalos*, ut in *Leucoio*, *Brassica*, *Thlaspi*, &c.
    { *Pentapetalos*, ut in *Lychnide*, *Caryophyllo*, *Alsine*, &c.
    { *Hexapetalos*, ut in Bulbosis. *Polypetalos* in aliis.

   Difformis, ut in *Viola*, *Papilionaceis*, &c.

 Compositus, qui ex pluribus flosculis, quorum singuli singulis insident seminibus, in unum totalem florem coeuntibus constat; éstque vel

  Discoides, in quo flosculi breves, arctè compressi unam quasi planam superficiem componunt, ut in *Calendula*, &c. est vel

   { *Radiatus*, limbo vel margine foliorum planorum discum cingente; foliis marginalibus vel *frontatis*, fronte *crenatâ*, ut in *Calendula* & pappossis lactescentibus, *æqualis* : vel *cuspidatis*, ut *Ptarmica Austriaca* Clus.
   { *Nudus*, qui petalis illis seu flosculis marginalibus caret, ut in *Tanaceto*, &c.

  Naturâ plenus, ut in Pappossis lactescentibus.
  Fistularis, ut in Capitatis dictis, *Jacea*, *Carduo*, &c.

Imperfectus, qui harum partium aliqua caret.

Flos est vel

Figure 6.5 *A chart of taxonomic distinctions to be used for classifying plants regardless of their places of origin. From John Ray,* Historia plantarum *(1686).*

the elaborate systems of such as John Ray in England, or Joseph Pitton de Tournefort in France – and by the 1730s to the attempts of the Swede Carl Linnaeus. But more was involved in the impact of the New World upon botanical practices than the increase in the number of things to be known. First of all, it is important to notice that the very means of collection had changed: it was in sixteenth-century Italy that Aldrovandi and other botanists first began to collect actual *specimens* of plants, rather than simply describing them *in situ*. This was essential to the notion of natural historical knowledge as being centred on collections of specimens brought from many different locations. It also shaped the classification of plants: names could be arranged even more easily than specimens into taxonomic systems that were *universal*; that applied everywhere.

There was also a psychic impact. Comparisons between the extension of natural knowledge and the growth of geographical knowledge, of the sort often made by Francis Bacon, transformed perceptions of the natural world. In the 1630s the Dutch diplomat Constantijn Huygens, father of the physico-mathematician Christiaan Huygens, wrote the following concerning the use of magnifying lenses:

> And discerning everything with our eyes as if we were touching it with our hands, we wander through a world of tiny creatures, till now unknown, as if it were a newly discovered continent of our globe.[26]

This is a conception of the natural world as a vast field for investigation, in which horizons enlarge the further we go. The institutional results of such a perception may be seen in a variety of European countries. Bacon's own vision of natural philosophy as discovery had started, as we have seen, in the 1590s with his proposal to Queen Elizabeth for the establishment of a botanical garden, amongst other things, as part of a national research enterprise.[27] In Italy, the Accademia dei *Lincei* bore a name that stressed much the same perceptual image as that expressed, more literally, by Constantijn Huygens.

The opening and expansion of the world that was associated with the geographical discoveries of the fifteenth and sixteenth centuries thereby stimulated the development of institutions dedicated to the pursuit of cumulative knowledge of the natural world: in a word, to the notion of *research*. Research implied the existence of things to be *found out*; and those things, as Francis Bacon was foremost in asserting, were matters of practical value. That practical value might, as Bacon sometimes suggested, have as its greatest virtue its capacity to underwrite truth; but it was also of great importance in its own right. Issues of trade, along with improvements in industry and agriculture, were at the heart of the attempted reforms made in England during the middle decades of the seventeenth century by the self-styled "Baconian" groups that were echoed after 1660 by the new Royal Society.

Research was also intimately connected to the idea of discovery, and here we see an important and specific sense of that word: "discovery" meant not just the finding out of something previously unknown, whether a new land or a new form of mathematical analysis (to both of which the term was applied); it meant taking that piece of knowledge and integrating it into a *system* – the common storehouse of European knowledge – that would enable its effective exploitation. That storehouse was not, however, truly common to all, because an extraordinary degree of organization was necessary to command it. By far the most important of the formal institutions that sought this goal were national governments and their agents. Although the Royal Society and, especially, the Academy of Sciences in Paris were among the first governmentally established bodies specifically devoted to "scientific" concerns, European expansion from the late fifteenth century onwards had already been intimately related to scientific matters. Those matters had simply been cloaked under other names, such as trade, diplomacy, and colonization. All participated in the same European global networks.

Chapter Seven
Experiment: How to Learn Things about Nature in the Seventeenth Century

I Reconfiguring experience

Aristotle had asserted unequivocally that all knowledge has its origins in experience. He was echoed by scholastic Aristotelians, so that the aphorism "there is nothing in the mind which was not first in the senses" became a standard philosophical maxim in the later Middle Ages.[1] Despite this fact, many non-Aristotelian philosophers in the seventeenth century had taken to criticizing the approaches to learning about nature that were promulgated by scholastic learning for *ignoring* the lessons of the senses. Francis Bacon was but one among many in his stated view that Aristotle "did not properly consult experience . . .; after making his decisions arbitrarily, he parades experience around, distorted to suit his opinions, a captive."[2] Bacon's became a common view: Aristotelian philosophy was commonly represented during the century as being obsessed with logic and verbal subtleties, reluctant to grapple with things themselves as encountered through the senses. The rhetoric of the Baconian Royal Society came equally to incorporate such a picture of Aristotelianism, its spokesmen making frequent remarks dismissive of scholastic obsession with words instead of things.

Galileo too, among many others, had attempted to dramatize what he saw as the emptiness of the official school philosophy. In Galileo's *Dialogo* of 1632, Simplicio (the Aristotelian character) at one point purports to explain why bodies fall by reference to their *gravity*. Salviati, who speaks for Galileo, replies by ridiculing the use of a *word* as an explanation. What is it that moves earthly things downwards? "The cause of this effect," says Simplicio, "is well known; everybody is aware that it is gravity." "You are wrong, Simplicio; what you ought to say is that everyone knows that it is called 'gravity.' What I am asking you for is not the name of the thing, but its essence, of which essence you know not a bit more than you know about the essence of whatever moves the stars around."[3]

Why was Aristotle's natural philosophy associated by its critics with a neglect of the lessons of experience and the favouring of empty words? The answers to this question will illuminate just what the new emphasis on experimental knowledge meant in the seventeenth century. As we saw in Chapter 1, section I, Aristotle's philosophy was centrally about understanding rather than discovery. Aristotle, while in practice very interested in empirical facts of all kinds (as found especially in his zoological writings), wanted above all to solve the problem of how we are to understand ourselves and the world around us. Thus, in his more abstract philosophical writings, such as the *Metaphysics*, or in his logical writings, the specific lessons of the senses are largely sidelined in favour of analyses of how to argue, how to understand, and in what terms we must make sense of our experiences. In the *Posterior Analytics* especially, Aristotle attempts to show how an ideal science should be structured so that it would be able to account for empirical truths; the acquisition of those truths was not centrally at issue, and neither were any particular such truths themselves. Thus when Aristotle's followers considered what Aristotelian natural science should look like, the model that they examined was one in which empirically acquired truths were taken as given, with only their explanation being the truly important task. In a sense, therefore, an Aristotelian world was not one in which there were countless new things to be discovered; instead, it was one in which there were countless things, mostly already known, left to be *explained*.[4] That Aristotle himself does not seem to have believed this is beside the point; it was nonetheless the lesson that his scholastic followers in medieval and early-modern Europe tended to draw from those of his writings that they found most interesting and most teachable.

The typical expression of empirical fact for such an Aristotelian was one that summed up some aspect of how the world works. "Heavy bodies fall" is a typical example: it was a statement that acted as an unquestioned reference-point in a network of explanations that involved such things as the terrestrial elements and their natural motions, final causes, and the structure of the cosmos.[5] Such statements appeared in already generalized form, rather than in the form of singular experiences referring to historically specific events. One did not say "this heavy body fell when I dropped it"; one simply said that all heavy bodies always fall – that is how nature behaves. In the absence of the reported particular, no room was left for the denial or affirmation of a universal claim about how *all* heavy bodies behave. The assumption was that everyone, from everyday experience, already knows it to be true. The philosopher's job, according to Aristotle, was to show *why* it was true. This was a matter of giving appropriate causal explanations that would, in the ideal case, show why the fact to be explained was *necessarily* true given the attendant circumstances. Needless to say, ideal cases were seldom, if ever, met with.

Understanding the sway, in early-modern Europe, of Aristotelian ways of formulating such questions involves seeing how even the most strongly

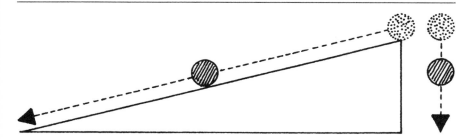

Figure 7.1 *Galileo's use of the inclined plane to slow down the acceleration of free-fall, thus making it easier to measure.*

anti-scholastic of philosophers could still take those ways for granted, as foundational aspects of their thought. For example, the dominant scholastic-Aristotelian way of conceptualizing and handling experience forms the backdrop to Galileo's famous work on the fall of heavy bodies, finally published in the *Discorsi* of 1638 although reflecting work largely completed by 1609.[6] Galileo tries at one point to establish the truth of his claimed experience that a falling body accelerates as it descends, its distance from the place of release increasing in direct proportion to the time elapsed. This experience takes the form of a standard Aristotelian generalization, describing how things behave in nature; Galileo does not describe a specific experiment or set of experiments carried out at a particular time, together with a detailed quantitative record of the outcomes. Instead, he simply says that, using apparatus of a kind carefully specified, he had found that the results of rolling balls down an incline and timing their passage yielded results that agreed exactly with his expectations, in trials repeated "a full hundred times." This last phrase (found frequently, in various forms, in contemporary scholastic writings) means, in effect, "countless times." Galileo wished to persuade his readers that the results amounted to common experience. His problem, however, was that the particular experience that he wished his readers to accept was *not* in fact one that is well known and familiar.

The subsequent rise to dominance of reported experimental events as the foundations of scientific arguments would be attended by just these difficulties. When a natural phenomenon was well known, it could be adduced as part of natural philosophical reasoning with no difficulty, because no one would be likely to contest it. But if the phenomenon were not well known, and instead brought to light only through careful and unusual experimentation, how could the natural philosopher make it acceptable for use in creating philosophical explanations? Galileo wished to have his readers believe that things behaved in nature just as he said they did. He could not rely on his readers already being disposed to accept the truth of

the foundational natural behaviours that he discussed (uniform acceleration in fall), but at the same time he could not allow the matter to rest on
nothing more than his say-so. Some people might have been prepared to
accept his claims on the basis of his own personal and institutional authority, but that would not have made his arguments *scientific*. Galileo always
adhered to a model of scientific demonstration that came straight from
Aristotle: a true scientific explanation should be demonstrative, like the
proofs of mathematics, and, like the mathematical theorems of Euclid,
proceed on the basis of simple statements that all could accept as true at
the outset. Euclid had employed starting points such as "when equals are
subtracted from equals, the remainders are equal"; they were intended to
be so intuitively obvious that no one could in good conscience deny them.
When Aristotelian natural philosophers made arguments on the basis of
empirical principles, such as "the sun rises in the east," or "heavy bodies
fall," they too relied on the practical undeniability of such truths; everyone
could be relied upon to accept them.[7] Experimental results, however, lacked
that kind of obviousness, which is why Galileo attempted, in the present
case, to render them as routine as possible as quickly as possible. Claiming
results that accrued from trials repeated "a full hundred times" was a way
of saying "things *always* behave in this way," and hoping that the reader
would believe it.

René Descartes confronted similar problems. Like Galileo, Descartes
finessed the problem of trust by refusing to acknowledge it as an issue. In
the *Discourse on the Method* (1637), he invites other people to assist in his
work by contributing "towards the expenses of the observations [*expéri-
ences*, which also means "experiments"] that he would need."[8] It was precisely the fecundity of his explanatory principles that required experiments,
because, as Descartes himself said, for any given natural phenomenon he
could usually imagine more than one possible explanation. Experiments
were therefore required to determine which of them might be the true
one. Descartes wanted to do all the actual work himself because, he says,
receiving information about phenomena from other people would typically
yield only prejudiced or confused accounts. He wanted to make the requisite experiences himself or else pay artisans to do them (since the incentive
of financial gain would ensure that the artisans would do exactly what
they were told). Descartes was intent only on convincing himself. He
sidestepped the problem of trust by adopting a supreme selfishness: what
convinced him should be good enough for anyone and everyone.

II Mathematical experimentation

These were issues that needed especial confrontation in the mathematical
sciences. As various kinds of "physico-mathematics" sprang up in the
course of the seventeenth century, the methodological impetus that had
driven the emergence of the category served also to emphasize difficulties

relating to experimental procedures.[9] The mixed mathematical sciences had often, since their ancient inception, involved the use of specially made apparatus to investigate natural behaviours that were not obvious from everyday experience. Thus astronomy used specialized sighting instruments for measuring precise positions of bodies in the heavens (well before the appearance of the telescope, an additional instrumental resource, in the seventeenth century). Optics used special devices for measuring angles in reflection and refraction. Ptolemy had written important treatises, the *Almagest* and the *Optics*, in both sciences, and he detailed the apparatus that was required for the proper conduct of work in each. The eleventh-century Islamic philosopher known in Latin Europe as Alhazen had written the most important optical treatise used in Europe prior to Kepler's studies, and he too detailed the makeup and use of optical apparatus.[10] As a result, the tradition of mathematical sciences practised by seventeenth-century Europeans involved them by its very nature in questions concerning the validation of artificially generated experience – experience that was *not generally known*.

Consequently, the ideal of an Aristotelian science, wherein the phenomena to be explained were taken as established from the outset, did not in these cases apply. The issue became especially pressing by the beginning of the seventeenth century among people such as the Jesuit mathematicians, who wanted to show that the mathematical disciplines were genuine sciences according to Aristotelian criteria (like Galileo, they were concerned about their status as mathematicians *vis-à-vis* the natural philosophers). Experimental apparatus gave them trouble because of its unobviousness.

Galileo's was a popular solution to this problem among mathematicians. Thus Jesuit mathematical scientists, such as the astronomer Giambattista Riccioli, reported experiments that involved dropping weights from the tops of church towers to determine their acceleration. While, unlike Galileo, Riccioli gave places, dates, and names of witnesses to underwrite his narratives, the way he used those narratives was always to turn them into authoritative assertions of how such matters *always turn out*. Another, especially famous, example of this presentational trick took place in 1648. The mathematician Blaise Pascal, perhaps best known for the famous "Pascal's Triangle," wrote from Paris to his brother-in-law, Florin Périer, in the Auvergne district of provincial France, requesting him to carry out an experiment. Pascal asked him to carry a mercury barometer up a nearby mountain, the Puy-de-Dôme, in order to see whether the mercury's height in the glass tube would change as the trial was conducted at different altitudes. Pascal hoped and expected that it would, because he was convinced that it was the pressure of the air that sustains the column of mercury in the tube, and that air-pressure decreases the higher one goes.[11] The apparatus was itself novel, having been devised in the 1640s in Florence by Evangelista Torricelli, who had been a protégé of Galileo's in the latter's

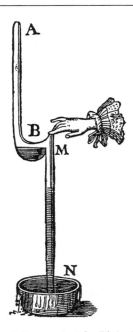

Figure 7.2 *Torricelli's experiment, in a variant by Blaise Pascal. The double arrangement is intended to demonstrate that the mercury is indeed supported by the pressure of the air.*

last years. Like Pascal, Torricelli ascribed the phenomenon to the weight, or pressure of the air (disputes also existed over which of the two, weight or pressure, was the correct way to speak of these matters).

Pascal published a narrative account of the experiment not long afterwards, a report written by Périer with Pascal's introduction and commentary. Périer provides a detailed account of his ascent and descent of the mountain, in the company of named witnesses, and records the height of the mercury that was found each time the apparatus was set up at various stops along the way. At the end of the story, which indeed showed that the mercury stood lower in the tube the higher up the mountain it was measured, Pascal proceeds to turns Périer's narrative into the keystone of a universal philosophical truth. First of all, Pascal uses Périer's results to produce a quantitative correlation of change in height of mercury with change in altitude, already taking it for granted that what Périer had recorded held true of all such measurements. Pascal then predicts the smaller changes in mercury height to be expected if similar apparatus were to be lifted up from the ground to the much lower elevations provided by church towers found in Paris – a more everyday setting than that of Périer's elaborate exploit. Finally, having made specific numerical predictions of the changes to be expected, Pascal then asserts that actual trials confirm the predictions. Like Galileo with his inclined-plane experiments on falling bodies, Pascal gives no details or particularities of these ecclesiastical experiments; they just agree with expectations, as good natural regularities should.

The two central difficulties raised up by experimental procedures, that of establishing trust in experimental narratives and that of establishing universality, or representativeness, for specific experimental outcomes, thus demanded answers with especial urgency in the mathematical sciences because these sciences often sought out unusual or unobvious phenomena. Opinions differed on what would happen to the height of mercury in the glass tube at increasing altitude, before Pascal's brother-in-law ascended the Puy-de-Dôme in an attempt to answer the question – a question that did not already possess a generally accepted answer. The mathematical sciences (which subsumed the work of Pascal and others on mercury barometers) provided their practitioners with specialized knowledge that was hard to use as the basis for a demonstrative science because it was not rooted in universally accepted experience. Somehow, therefore, specialized knowledge had to be *made* into common knowledge. A frequent recourse for astronomers and other kinds of mathematicians was to rely on their individual reputations as reliable truth-tellers. In many cases (such as that of the Jesuit mathematicians), corporate reputations could also be drawn upon: professorships in universities and colleges, or, as in Galileo's case, association with powerful sources of patronage, could lend subtle weight to empirical claims: challenge the result and you were challenging the institution that implicitly certified it.

Astronomers, however, had additional, more concrete ways of bolstering their claims. This is because, traditionally, astronomers did not as a rule publish their raw astronomical data. They did not present lists of observational results, such as measurements of planetary positions, which would then have required acceptance based solely on the astronomer's authority (unless, extraordinarily, similar measurements had been made by others at exactly the same times).[12] Instead, astronomers used their raw data to generate predictive tables of planetary, solar, or lunar positions, using geometrical models designed to mimic apparent celestial motions. This work was presented in such a way as to efface any formal distinction between *observational* astronomy (writing down the numbers that were measured using observational instruments) and those parts of the enterprise centred on the *calculation* of predictive tables from geometrical models – models that were themselves initially justified by their correspondence to the data.

This latter work was the part that might be deemed suitable for publication, but not the former. The predictive tables, rather than the original raw data, served as the *public* warrant for the goodness of the models from which they were computed, since anyone could check at any time to see how accurate those predictions were. In the sixteenth century, after all, Nicolaus Copernicus's reputation as an astronomer rested on his mathematical abilities, not his presumed competence as an observer; astronomers were *mathematicians*. Later on in the century, Tycho Brahe, although famous as an indefatigable observer, did not publish his vast accumulation of observational results; instead, he published mathematical treatments, employing his observational data, of such things as the paths of comets, or of his new earth-centred astronomical system. Tycho hired Johannes Kepler to compute a more accurate model for the motion of Mars on the basis of his raw data, without at the same time allowing Kepler free access to his complete observational records. These records were so far from public that Kepler himself had great difficulty in gaining control of them from Tycho's widow following Tycho's death.

"Experimentation" in the mathematical sciences, then, called on problems related both to trust and to the meaning of results relating to specific times and places. Astronomical practice already addressed such difficulties, as well as potential problems relating to the use of instrumentation in gathering data. In the latter case, instrumentation and apparatus, while usual for the mathematical sciences, were more problematic for areas of inquiry related to qualitative sciences. Francis Bacon's refusal to accept the legitimacy of a distinction between natural and artificial processes (as processes produced with artificial apparatus would be) thus plays an important role in the rhetoric, logic, and practice of experimental science in the seventeenth century.[13]

III "Baconian" experimentation

As we saw in the previous chapter, Bacon's writings were used as an important resource for justifying experimental investigations, especially by the Royal Society of London. Bacon's own position on experiment as a scientific tool is, however, more ambiguous than it at first appears.

Bacon, like Aristotle, stressed the importance of experience in learning the ways of nature. The examples that Bacon used to illustrate a proper use of deliberately contrived experience in making (his kind of) natural philosophical knowledge show exactly the same features of generality, or universality, that characterize the writings of scholastic philosophers. In Book II of the *New Organon* (1620), Bacon presents two worked examples of his new logic of investigation (usually referred to as his "method," although he never called it that). One of these examples concerns the nature of heat: among the listed "Instances meeting in the nature of heat" we find "the sun's rays, especially in summer and at noon"; "solids on fire"; "quicklime sprinkled with water"; and "horse shit, and similar excrement, when fresh."[14] Notice how every one of these is an assertion of a general truth applying to every case of each "instance"; Bacon evidently sees no need to adduce specific observations. This (in its own context, unremarkable) habit is seen again when he refers to some instances of variation in the degrees of heat found in varying circumstances. In giving examples, Bacon sometimes proposes tests the outcomes of which he already knows:

> Try an experiment with burning glasses in which (as I recall) the following happens: if a burning glass is placed (for example) at a distance of a span [i.e. nine inches] from a combustible object, it does not burn or consume as much as if it is placed at a distance of (for example) a half-span, and is slowly and by degrees withdrawn to the distance of a span. The cone and the focus of the rays are the same, but the actual motion intensifies the effect of the heat.[15]

The universality of this description of an experiment is part of its very effectiveness. By describing a trial the outcome of which Bacon claims to know, from the warrant of personal experience ("as I recall"), he tells the reader about something that happens in nature without actually tying it down to a specific event, an occasion on which this was tried with this outcome. Presenting experience in such a manner served to bypass, at least rhetorically, the difficulties that would arise if Bacon's argument had depended on taking his word for an historical event that lacked corroborating witnesses (recall, too, that Bacon was a lawyer). By telling you what *happens* rather than what *happened*, and by giving an account in the form of instructions as to what to do to produce this claimed effect, Bacon can create the illusion of having revealed to his reader a fact about the natural world, one that can then be used to undergird a philosophical argument about the nature of heat.

The form of "Baconianism" adopted, or asserted, or claimed, by the Fellows of the early Royal Society was one centred on the notion of *utility* rather than of *experiment*. Although the early Royal Society is often regarded as a bastion of experimentalism, the kind of experimentalism that it practised was different from that of Bacon, in the same way that it was different from Aristotle's. Where the hallmark of Aristotle's, or Bacon's, kind of scientific experience was the universal generalization, the attempt to appeal to common experience, the hallmark of the Royal Society's was the particular event. When a Fellow of the Royal Society told his audience about an experiment, he did not usually provide a recipe that purportedly revealed a regular feature of the world, as Bacon might have done. Instead, he typically told a story about an event that had happened in the past, to him, at a specific time and place. He did not, that is, make an immediate jump from a particular personal experience to an account of how some aspect of nature habitually behaves.

Here is a quite typical example from the writings of Robert Boyle:

We took an open-mouthed glass, such as some call jars, and ladies often use to keep sweetmeats in, which was three inches and a half, or better in diameter, and somewhat less in depth, and had the figure of its cavity cylindrical enough. Into this having put some water to cover the protuberance wont to be at the bottom of such glasses, we took a convenient quantity of bees-wax, and having just melted it, we poured it cautiously into the glass, warmed before-hand to prevent its cracking, till it reached to a convenient height.[16]

And so the account continues, circumstantially and with considerable detail, describing an experiment that was intended to refute some criticisms levelled against Boyle's earlier experimental work by Henry More. Boyle's exposition concludes in similar style: "And lastly, we took off by degrees the grain weights that we had put on, till we saw the wax, notwithstanding the adhering lead, rise, by degrees, to the top of the water, above which some part of it was visibly extant."[17]

This style is quite standard for Royal Society publications, including articles in its unofficial journal, the *Philosophical Transactions*. The style went along with a determination on the part of the Fellows to steer clear of speculation or hypothesis, in favour of reporting solid facts. The purpose of such an ethic was not to prevent anyone from making conjectures about natural phenomena and their causes, but to avoid the appearance of a dogmatic adherence to any particular hypothesis on the part of the Society itself. Thus the Society's Curator of Experiments, Robert Hooke, wrote to the Society at the start of his *Micrographia* (1665), that in the book

there may perhaps be some *Expressions*, which may seem more *positive* then [sic] YOUR Prescriptions will permit: And though I desire to have them understood only as *Conjectures* and *Quæries* (which YOUR Method does not altogether disallow) yet if even in those I have exceeded, 'tis fit that I should declare, that it was not done by YOUR directions.[18]

And like Hooke himself, Boyle and other Fellows typically couched such cautious explanations in the terms of corpuscles and their behaviour.

The Royal Society used talk of a Baconian eschewal of hypotheses (which Bacon had decried as "Anticipations of Nature") to retain the integrity of its enterprise: their work was to rely on building up solid accumulations of facts. For this purpose, the particularities of reported, historical experiments, with no positive guarantee that attempted replications would be successful, were the simplest and safest things to discuss. The work of building up reliable theories to subsume and explain those facts was not thereby abandoned, but Boyle and others often spoke of that following stage of their work as residing in the future, to be tackled only when enough solid facts had been accumulated.

The approach of the Royal Society was not to the liking of all natural philosophers in this period, even in England. One of the fiercest critics of the Society was the philosopher Thomas Hobbes, later best known for his political philosophy. Hobbes had served as a secretary to Francis Bacon towards the end of the latter's life, yet despite this personal history, he was dismissive and scornful of the kind of "experimental philosophy" advocated and practised by Robert Boyle and his kind. Hobbes's reasons for this came out most strongly in his critique of Boyle's experiments with air-pumps, in which Boyle had conducted and written about the behaviour and properties of the space left inside an air-pump "receiver" (the glass globe from which the air was pumped). Hobbes poured scorn on Boyle's contention that he had, in these trials, removed practically all the air from the receiver, and in so doing, Hobbes also denigrated the value of such experimental investigation in general.

Hobbes's central objection was that the performance of experiments was not *philosophical*. Knowledge about nature was supposed to be natural philosophy, after all, and yet the kind of knowledge proposed by Boyle and others failed to achieve the universality and necessity that true scientific explanations by definition required. In this, in other words, Hobbes remained wedded to the Aristotelian understanding of what made a true science. Boyle spoke about experiments as historical events, whereas Hobbes wanted to produce demonstrations that would prove their conclusions with necessity, like mathematical demonstrations. Furthermore, Boyle's air-pump experiments consisted of trials conducted using complicated apparatus; why, Hobbes wanted to know, would you examine the behaviour of complex situations before you could make sense of simple, everyday ones?

Boyle emphasized experiment as the best way to make knowledge of nature that would command general assent. Everyone would be able to see for themselves that what was claimed was actually true. Hobbes objected that the kind of knowledge that this represented failed to yield *explanations* for natural phenomena. At best, Boyle could display natural behaviours to which everyone might assent, but there was no way in experimental work to demonstrate what the causes of those behaviours must be. Hobbes stressed the point that, whatever interpretation Boyle might provide for one of his phenomena, Hobbes could always come up with several different ones, each as likely as Boyle's. Hypothetical explanations were easy to make, but, for Hobbes, they were not sufficient to make a true natural philosophy, and he accused Boyle of asserting the existence of a vacuum (which Hobbes denied to be possible) on insufficient grounds:

The science of every subject is derived from a precognition of the causes, generation, and construction of the same; and consequently where the causes are known, there is place for demonstration, but not where the causes are to seek for. Geometry therefore is demonstrable, for the lines and figures from which we reason are drawn and described by ourselves;

and civil philosophy is demonstrable, because we make the common-wealth ourselves. But because of natural bodies we know not the con-struction, but seek it from the effects, there lies no demonstration of what the causes be we seek for, but only of what they may be.[19]

Consequently, for Hobbes, the best that could be done in natural philoso-phy was to postulate possible causes (he favoured mechanical ones) that were capable of explaining the observed phenomena; but the truth of those causes could never be demonstrated.

Boyle, like most of the leading Fellows of the Royal Society, was himself cautious about hypotheses. His care to avoid dogmatic talk or to ascribe causal explanations in his work led him, for example, to refuse to speak positively on whether the action of the air-pump created a true vacuum in the receiver; that is, whether the space became truly empty. Instead, he spoke of the removal of the "ordinary air," leaving open the possibility that there might be some weightless, undetectable, aetherial medium still present. Boyle used the word "vacuum" to refer to the space inside the receiver when once it was emptied of air, but he made it clear that this operational vacuum was not to be confused with a "metaphysical," true vacuum. Whether a true vacuum existed was a question on which he refused to pronounce, Hobbes's charges to the contrary notwithstanding.

Furthermore, Hobbes's own infatuation with the mathematical, demon-strative model of science was not one from which Boyle radically departed, insofar as this generally accepted ideal could be applied. As he wrote regarding work on buoyancy and displacement, "it is manifested by hydro-staticians after *Archimedes*, that in water, those parts that are most pressed, will thrust out of place those that are less pressed; which both agrees with the common apprehension of men, and might, if needful, be confirmed by experiments."[20] Thus, in establishing for practical purposes the truth of this hydrostatical principle, Boyle was as ready to use "the common appre-hension of men" as his starting point as was Aristotle, or Euclid. Experi-mental confirmation was simply something that was available "if needful." But in matters that were novel and unobvious, special experimental con-trivances and their disciplined management were central to Boyle's view of how to learn things about nature.[21]

The *Saggi* of the Accademia del Cimento, published in 1667, were subsequently translated into English by another Fellow of the Royal Society, Richard Waller, and published in 1684 as *Essayes of Natural Experi-ments*. The anonymity and recipe-like generality of many of the *Saggi's* experimental accounts are somewhat reminiscent of the impersonal recipes by which instruments and their proper uses were often described in mathematical treatises of astronomy or optics; but the first-person (albeit unnamed), circumstantial accounts of the conduct of very many of the experiments suited perfectly the model adhered to by the Royal Society. For example:

To throw some light on the question, whether the cooling of a body results from the entry of some kind of special atoms of cold, just as it is believed that it is heated by atoms of fire, we had two equal glass flasks made, with their necks drawn out extremely fine. These were sealed with the flame, and we placed one in ice and the other in hot water, where we let them stand for some time. Then, breaking the neck of each under water, we observed that a superabundance of matter had penetrated the hot one, blowing vigorously out of the flask. . . . It seemed to some of us that the same thing should have occurred when the cold one was opened, should the cooling of the air in it have proceeded in the same way . . . i.e., by the intrusion or packing in of cold atoms blown by the ice through the invisible passages of the glass. But it turned out quite the other way.[22]

The centrality of experiments and experimental reports to the business of the early Royal Society resonates awkwardly, therefore, with the work of one of the Society's most celebrated members, Isaac Newton. Newton was a university mathematician, from 1669 the successor to Isaac Barrow as Lucasian Professor of Mathematics in the University of Cambridge, and a man who first came to the Society's collective attention in 1671. He was already familiar with the Royal Society and its work, having studied, among other things, volumes of the *Philosophical Transactions* in the 1660s. Newton evidently wanted to become associated with the group, and to that end sent them a small reflecting telescope of his own design and manufacture. The Fellows rewarded the young Cambridge mathematician with an election to the fellowship. Encouraged, Newton soon afterwards sent to Henry Oldenburg, in the latter's guise as the Society's secretary, a letter describing for the Royal Society some of his studies on optics that related to the ideas behind the telescope that he had sent them.

This letter was not long after published in the *Philosophical Transactions* as "A Letter of Mr. Isaac Newton, Professor of Mathematics in the University of Cambridge; Containing His New Theory About Light and Colours."[23] One of the many features of this celebrated paper is its use of a particularistic, event-focused experimental format to present material that would normally have fallen under the heading of the mathematical science of optics. Thus Newton begins by telling a story about events that had transpired back in 1666. He tells of how he had, for no good reason, got himself a glass prism, and used it to cast a spectrum generated from the rays of the sun projected through a hole in the shutters of a darkened room. (Newton was not the first to play with prisms in an optical investigation; Descartes had used one in his essay "Dioptrics," for instance.) He says that he was "surprised" by the oblong shape of the spectral band of colours, "which according to the received laws of refraction, I expected would have been circular."[24] The length of the spectrum was, he says, five times its breadth, "a disproportion so extravagant, that it excited me to a more than ordinary curiosity of examining from whence it might proceed."[25] Newton's historical account of what happened, and what he did, leads the reader towards a general conclusion that the light from the sun spreads out into a band when refracted through a prism because it is composed of "difform rays, some of which are more refrangible [i.e. "able to be refracted"] than others: so that of those, which are alike incident on the same medium, some shall be more refracted than others, and that not by any virtue of the glass, or other external cause, but from a predisposition, which every particular ray has to suffer a particular degree of refraction."[26]

Furthermore, Newton proceeds to assert, those differing degrees of refrangibility correspond to differing colours of the light exhibiting them. Those rays which are refracted most exhibit the blue-violet colour characteristic of one extreme of the spectrum, whereas those rays which are refracted the least correspond to the red colour visible at the opposite end of the spectrum. The refrangibility of each kind of ray is an unalterable property, remaining constant throughout a number of successive refractions and reflections; furthermore, the colour associated with any particular refrangibility of ray is similarly unalterable. Thus Newton could ascribe numbers to colours, by characterizing any spectral colour in terms of the degree of refrangibility of its ray.

Newton's optical paper to the Royal Society thus goes out of its way to appear non-mathematical. Newton does not provide a geometrical diagram to assist in his preliminary exposition of these experiments; instead, he presents the first part of the paper as an experiment of the kind he knew the Royal Society preferred, an historical account of what he had seen and done on a particular occasion in the past. A shift to a more typical mathematical format, in which general conclusions are stated, occurs only after the central experimental premises have been laid out in narrative form. Fittingly, Newton incorporates into his letter remarks regarding the problems caused by the differential refrangibility of light rays for making telescopes that will focus light-sources precisely rather than blurring them, and explains how he had come to make his reflecting, instead of refracting, telescope as a consequence. The practical, operational, Baconian dimension of the new experimental philosophy was an important part of Newton's enterprise.

Newton's own work came to represent a conception of scientific experience that departed considerably from the old scholastic model, therefore. For an Aristotelian philosopher, "experience" was the proper source of knowledge about the world's habitual behaviour. For Newton and his later followers (and see Chapter 8, below), experimental philosophy was now a means for interrogating nature that yielded, above all, operational rather than essential knowledge – it told you how to do things, rather than what something truly was in itself. Experimentation, as the Royal Society understood it and as Newton refined it, became an approach to knowledge that

accumulated records of natural phenomena that owed their general credibility to institutional authority or to the word of appropriate witnesses
(Boyle's especial technique).

IV Physiological experimentation

William Harvey's investigations show once more the importance of the
accepted, broadly Aristotelian framework for experimental studies in this
period, as well as the specific difficulties of experimental study in physiology. His work also further indicates the kinds of practical means available
for dealing with problems of credibility.

Harvey's *De motu cordis* of 1628 had opened, as we saw in the previous
chapter, with two dedicatory prefaces, one to the king, the other to the
College of Physicians. The latter preface did some important work for
Harvey, because he was proposing a view of the behaviour of the heart and
blood that flew in the face of long-accepted Galenic teaching. Galen (like
Aristotle) had taught that the heart is a kind of repository for the blood,
which is communicated out to the rest of the body through the network of
blood vessels. Galen's specific version of this view distinguished between
the system of the arteries, branching out from the left side of the heart, and
the system of the veins, which connected to the right side of the heart but
was regarded as having its "origin" in the liver. Arterial blood carried heat
and *pneuma* (a kind of vitality derived from the air in the lungs) out from
the heart to all parts of the body. The veins had a different function, that of
distributing nutrition around the body. Venous blood was created in the
liver from ingested food, which is why the veins were seen as having their
origin in the liver. Blood found its way into the arterial system, where it
served its quite different distributive function, by seepage through pores
in the wall of the heart. This wall, called the septum, divided the left side
of the heart from the right, and the pores in the septum were the only means
of communication between the one side and the other that Galen could
imagine. The beating of the heart helped in expressing blood out from the
heart, but there was no circulatory pumping.

Harvey, by contrast, saw the arterial and venous systems as two components of a larger circulatory system. Blood was pumped out from the
left side of the heart through the arteries. The arteries, as they are traced
out by the anatomist from the heart, branch out and become, as they do so,
more numerous, smaller, and finer. Harvey held that the ultimate status of
these branching arteries was as invisibly small blood vessels that gradually
linked together again to form the venous system, which served to return
the blood to the right side of the heart. So blood left the heart through
the arteries and returned to the heart through the veins. Furthermore, there
were no pores in the septum. Instead, venous blood found its way to the
heart's left side by making a "pulmonary transit" from the heart's right
side through special blood vessels that carried it through the soft, spongy

tissue of the lungs (with the blood vessels again having subdivided into invisible tubelets), before returning through appropriate blood vessels from the lungs to the left side of the heart. The full circulation then having been completed, the blood could thereafter be sent out once again via the arteries.

The "pulmonary transit" was an idea that had already been put forward at Harvey's *alma mater*, the University of Padua, in the later decades of the sixteenth century, and was the element of his mature ideas that Harvey had presented in his 1616 Lumleian lectures.[27] The full, or "general," circulation around the body was Harvey's real, and spectacular, innovation.

Now, this picture was not one that could be demonstrated by simply opening a living animal body and looking. Its establishment required Harvey to make a large number of experiments on a wide variety of animals, from shellfish to human beings, and to elucidate what he saw by means of arguments. One of the chief difficulties of the work was in making it clear to others that he really had seen what he claimed to have seen, and that his inferences genuinely followed from that evidence. This is where the preface addressed to the College of Physicians played an important role:

> The booklet's [i.e., *De motu cordis'*] appearance under your aegis, excellent Doctors, makes me more hopeful about the possibility of an unmarred and unscathed outcome for it. For from your number I can name very many reliable witnesses of almost all those observations which I use either to assemble the truth or to refute errors; you so instanced have seen my dissections and have been wont to be conspicuous in attendance upon, and in full agreement with, my ocular demonstrations of those things for the reasonable acceptance of which I here again most strongly press.[28]

In effect, Harvey was informing potential critics that if they doubted or denied his assertions, they would at the same time be doubting or denying the "full agreement" of the members of the most illustrious medical institution in England. These sorts of social relationships, whether with a royal patron, a socially accredited professional society, or even with respected gentlemen, all served to render more plausible an individual's truth claims. Experimental assertions, in order to be treated as if they were philosophical assertions, needed as much shoring up as they could get, from whatever quarter available.

Harvey himself, when later debating his views on circulation with a critic, stressed the fundamental issue at stake: "Whoever wishes to know what is in question (whether it is perceptible and visible, or not) must either see for himself or be credited with belief in the experts, and he will be unable to learn or be taught with greater certainty by any other means."[29]

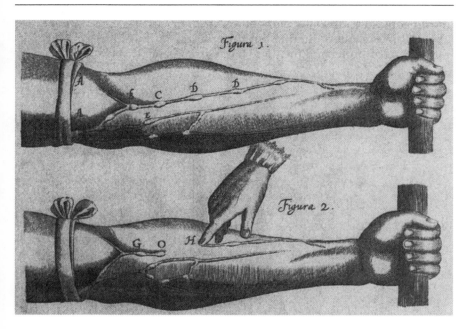

Figure 7.3 *An "ocular demonstration" of the function of the valves in the veins, from* Harvey's De motu cordis.

Harvey wanted this necessary recourse to experience and accredited testimony to be accepted as legitimate in making natural philosophy. To establish the point, he appealed to the usual touchstone of certain knowledge, mathematics: "If faith through sense were not extremely sure, and stabilized by reasoning (as geometers are wont to find in their constructions), we should certainly admit no science: for geometry is a reasonable demonstration about sensibles from non-sensibles. According to its example, things abstruse and remote from sense become better known from more obvious and more noteworthy appearances."[30] If mathematics can be accepted as certain and scientific, so too should a properly conducted experimental science – such as his own work in physiology.

The senses remained paramount in the sciences revolutionized by the new breed of philosopher in the seventeenth century, therefore, and one of the key tools for generating knowledge from them was the experiment. Experiment, understood as the making of specific trials of phenomena, typically with contrived circumstances or apparatus, was a particular kind of sensory experience that went beyond a simple inventory of what all or most people already knew was in the world. In this sense, experiment was about *discovery*, about finding out new things. As such, it had to incorporate means of protecting the discoverers from being disbelieved.

Chapter Eight
Cartesians and Newtonians

I Cartesian natural philosophy in France

"Cartesian" natural philosophy, as it became established and practised in the last decades of the seventeenth century, did not always follow closely the ambitions of its originator. Two of the most prominent adopters of Descartes's approach to physical explanation, Christiaan Huygens and Jacques Rohault, departed significantly from the master's conception of true natural philosophy as represented in such works as the *Discourse on the Method* or the *Principles of Philosophy*. They did so by emphasizing the *hypothetical* character of their explanatory mechanisms.

Huygens, as we saw in Chapter 6, became one of the leading figures in the Royal Academy of Sciences, in Paris, in the 1660s. He had first become fascinated by the physics of Descartes in the late 1640s, when, as a teenager, he had become attracted to Descartes's mathematical, or quasi-mathematical, approach to natural philosophy. Descartes was a personal acquaintance of Christiaan's father, the prominent diplomat Constantijn Huygens, and Christiaan's exposure to a form of Dutch Cartesianism had deep roots in his early life. His own response to such innovation was, however, more physico-mathematical than it was metaphysical. Descartes was centrally concerned with the establishment of a secure foundation for an account of the physical world that could rely on mathematical reasoning *without leaving anything out* (see Chapter 5, section III, above). But Huygens cared primarily about what mathematical and mechanistic approaches could be used for, and what they could achieve in the way of practical results. Thus one of his earliest forays into physico-mathematical work took the form of an analysis in 1646, when he was seventeen years old, which his proud father saw fit to send to Descartes's correspondent in Paris, Marin Mersenne, for Mersenne's evaluation. This was an examination of the implications of a particular model of gravity; one that saw its effects as being the cumulative result of a rapid succession of discrete impulses, for the acceleration of falling bodies. Christiaan had concluded that such a model should result in the steady, uniform acceleration of a falling body – just the result that Galileo (and, by this time, a good many others) had already suggested, although it is not clear whether Christiaan himself was aware of it at that time.[1]

During the 1650s, Huygens conducted much work (unpublished until 1703, after his death) on mechanics and motion, including the application of a formal principle of the relativity of motion to determining the outcome of collisions between perfectly elastic bodies (*De motu corporum ex percussione*, "On the Motion of Colliding Bodies," written in 1656). He also wrote during this same period another unpublished work called *De vi centrifuga*, "On Centrifugal Force," which analysed constrained motion about a centre and gave the apparent outward tendency of the revolving body the name that it has retained ever since. Circular, or, specifically, vortical motion was of particular interest within Cartesian physics, as we have seen.

Huygens's mature discussions of gravity, however, which were presented to the Academy of Sciences in 1668 and 1669, possessed much more explicitly hypothetical, conjectural characteristics than these earlier studies. They concerned a theory of gravity derived from a basic idea owing to Descartes, which attempted to explain terrestrial gravity in terms of submicroscopic particles revolving around the earth at enormous speeds and in all planes of rotation about its centre. Huygens was thereby able to calculate the necessary speeds of these particles such that the resultant gravitational acceleration towards the earth's centre exhibited by ordinary bodies (equal but opposite in force to the centrifugal force of the tiny particles themselves) would equal the empirically determined figure. In the subsequent published version of this discussion, the *Discourse on the Cause of Gravity*, Huygens writes as follows:

> I do not present [the hypothesis] as being free from all doubt, nor as something to which one could not make objections. It is too difficult to go that far in researches of this kind. I believe, however, that if the principal hypothesis upon which I ground myself is *not* the truthful one, there is little hope that one could find it [i.e. the correct hypothesis] while remaining within the limits of the true and sound philosophy.[2]

In exchanges about gravity among members of the Academy in 1669, Huygens had also expanded considerably on the hypothetical nature of the theory, its relation to Descartes's work on gravity, and what he meant by "the true and sound philosophy." His remarks show the extent to which Descartes's search for certainty had become, in Huygens's watered-down version, a search for *intelligibility* instead. That is, Huygens had become convinced that a mechanical philosophy of the kind advocated by Descartes, while not provable as the necessarily true account of the universe, was at least privileged as being the only one that could provide explanations that were truly intelligible; that is, explanations that made perfect sense. Thus Huygens presents his own "hypothesis" of the cause of gravity as one that accounted for the observed phenomena as well as (or, rather, better than) any other known account, and did so within the explanatory limits of a world-picture containing nothing but inert matter in motion. That world-picture defined what was and was not "intelligible." Huygens maintained that anything violating the explanatory limits of such a world-picture simply *would not make sense.*

This is how Huygens framed the issues in 1669, at the very beginning of his own paper:

> To find an intelligible cause of gravity, it's necessary to see how it can be done while supposing [i.e. postulating] in nature only bodies made of the same matter, in which no quality is considered, nor any inclination for each to approach the others, but only different sizes, shapes, and motions.[3]

Huygens goes on to note how, of these few admissible properties of material bodies, only motion seems to be suitable to explain an "inclination to motion" such as gravity. He continues by explicating his hypothesis, making frequent reference to Descartes's account of gravity, explaining those points on which his differed from Descartes's, and why. One part of his exposition involves citation of an experiment to display the inward tendency on bodies generated by a rotational movement of the fluid medium in which they swam; a two-dimensional parallel to Huygens's own three-dimensional theory, this illustration nonetheless exhibits an authentically Cartesian approach to clarifying theoretical accounts using mundane examples.[4] (See Figure 8.1.)

That Huygens always saw what he was doing in relation to Descartes's example is further illustrated by remarks in the Preface to the 1690 *Discourse*:

> Monsieur Descartes saw better than his predecessors that one could never understand anything more in physics than what could be referred to principles that do not exceed the limits of our mind, such as those that concern [*dependent des*] bodies (considered without qualities) and their motions. But since the greatest difficulty consists in showing how so many different things are produced from these principles alone, it's in that regard that he did not greatly succeed in several particular subjects that he set himself to examine: including, in my opinion, among others that of gravity.[5]

Huygens thus justified his own attack on the subject with reference to Descartes's project, which he saw himself as carrying forward.

Huygens's self-perception as a Cartesian cheerfully disregarded aspects

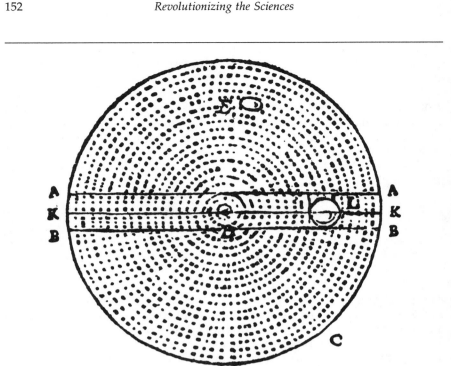

Figure 8.1 *Huygens's account of gravity in terms of a spinning fluid: the whole (horizontal) disc is rotated rapidly, and then halted. The fluid, however, continues to rotate, and the solid constrained body L is forced towards the centre of the vortex.*

of Descartes's thought that are nowadays regarded as central. Huygens represented Descartes as having embarked on a project in physics that was based upon "principles that do not exceed the limits of our mind," rather than principles that were metaphysically grounded and absolutely certain. Huygens sought "intelligibility" in his physical explanations, as did Descartes; Descartes had recognized that explanations of particular phenomena might not themselves be certain even if they were constructed on the basis of starting principles that were – hence the need in many cases for experiments. Descartes had used the metaphor of a watch to explicate the point: we know that a watch operates on the basis of cogs and wheels that serve to translate the action of the mainspring into the motion of the watch's hands, but if all we can see from the outside is the movement of those hands, while the arrangement and motions of the mechanism inside the watch-case remain hidden to us, we cannot tell precisely what that internal mechanism is like. This is because it is always possible to imagine a variety of different arrangements of cogs and wheels inside the watch-case, all of which would be capable of producing the same external movements of the hands. Such, thought Descartes, might also be the case for natural phenomena and their explanation in terms of invisible micromechanisms. Only carefully contrived experiments might serve to distinguish between various possible alternatives, even then not rendering the favoured alternative certain (since still more alternatives might typically lurk unnoticed or unexamined). However, Descartes based all of this reasoning on the assumption that the basic explanatory principles themselves (like the cogs and springs of a watch) were unquestionable; *any* explanation would need to be consistent with them, because they were the metaphysically certain foundations of all physical phenomena.

Huygens did not agree; or, he chose not to understand Descartes in this way. Instead, Huygens took Descartes's metaphysical arguments regarding such things as the nature of matter to describe the limits of the human mind only: human beings *cannot understand* explanations that are couched in terms other than those of inert matter in motion, and its behaviour, but there is no guarantee that that human inability reflects the true condition of reality. It could be that there are truer explanations of phenomena that will not serve us for explanations because they would make no sense to us. That is a human limitation, not a guarantee of how God chose to make the universe.

Huygens expressed methodological views of a similar sort to those concerning his explanation of gravity in another, similar study also written originally as a presentation to the Academy, read in 1679 and published in 1690. At the outset of this *Treatise on Light*, Huygens discusses his reasons for treating the behaviour of light as if it were a form of motion: "It is inconceivable to doubt," he says, "that light consists in the motion of some kind of matter." He describes the effects of heat, and its dissolution of ordinary matter by burning, because fire and flame are the means by which light is typically engendered here on earth. "This is assuredly the mark of motion, at least in the true Philosophy, in which one conceives the causes of all natural effects in terms of mechanical motions." But he does not say that this "true Philosophy" of necessity tells us the *truth*. Instead, he justifies the use of nothing but "mechanical motions" as one's explanatory principles as follows: "This, in my opinion, we must necessarily do, or else renounce all hopes of ever comprehending anything in Physics."[6] There is a right way of philosophizing which represents the best we can do; but it provides no guarantees. And we are, so to speak, stuck with those principles, because of their peculiar intelligibility as shown by Descartes.

The outcome of these considerations was an interpretation of light as consisting of longitudinal waves in a fluid, aetherial medium, much like sound-waves in air – the latter idea having become a conventional understanding of sound during the seventeenth century, following the work of, among others, Isaac Beeckman. Curiously, perhaps, Huygens did not consider that differing wavelengths might correspond to sensible differences in the appearance of light such as colours; instead, he put forward a view in which light was understood as a kind of jumble of impulses propagating through the fluid medium. It was the *speed* of the waves (finite, in contrast to Descartes's view of the instantaneous propagation of light-pressure), and the way in which spherically expanding waves combined to constitute a wavefront, that centrally concerned him. Wavelength was not a variable that was relevant to the problems that Huygens was concerned to address. This, indeed, may be the most useful way to understand Huygens's version of "Cartesian" mechanistic physics: he treated individual problems, individual phenomena, as matters to be addressed in piecemeal fashion, controlled only by the necessity to make sense of every one of them by the peculiarly intelligible Cartesian principles of "mechanical motions." A perfect consistency among all the individual models invented to deal with individual natural phenomena was not an immediate goal of the enterprise. Thus the subtle particles shooting around the earth that Huygens used to understand gravity did not have to be reconciled with the stationary medium that bore light waves.

Huygens may have been the more influential in serious natural-philosophical circles such as the Academy of Sciences, but Jacques Rohault was surely the major exponent of (a version of) Cartesian natural philosophy in educated Paris at large. Rohault became a prominent and successful public natural philosophy lecturer in the city in the 1660s, and promoted the ideas of Descartes in the context of applied mathematical and experimental demonstrations of physical phenomena, such as optical, barometric, and magnetic effects. In 1671 he published his *Traité de physique* ("Treatise of Physics"), a systematic discussion of the subject deriving from his lectures. In that work, Rohault stressed, conventionally enough, the importance of both reason and experience in creating knowledge about nature. He did so in a "Preface" that presents a discussion of the history of learning since antiquity, an approach that we have already seen used by Bacon. Like Bacon, Rohault emphasizes the way in which, as time goes on, knowledge grows, and that therefore too great a respect for the thinkers of antiquity, especially Aristotle, is misplaced. Echoing Descartes, Rohault mocks the supposed obscurity and unintelligibility of Aristotelian philosophical definitions (specifically, that of motion, precisely as Descartes had ridiculed it in *Le monde*). And echoing Bacon, Rohault also repeats (again without attribution) Bacon's dismissal, in the *New Organon*, of disputes on the divisibility of matter, saying, like Bacon, that such disputes are worthless because they have no *practical* implications at all.[7] This concatenation in Rohault of the supposed clarity of Cartesian explanatory principles with the operational criteria of Baconian natural philosophy is typical of Rohault's practical, unmetaphysical presentation of mechanistic explanations for experimentally produced phenomena. In addition, Rohault stresses the importance of mathematics for generating understanding in all manner of investigations, and in fact bemoans the usual tendency to separate mathematics from the rest of philosophy, criticizing "the Method of Philosophers" for "neglecting Mathematicks to that Degree, that the very first Elements thereof are not so much as taught in their Schools."[8]

The body of Rohault's text consists of a systematic description of the principal divisions of nature, namely the constituents of physics in general (matter, the true nature of the qualities apprehended by the senses, and so forth); cosmography (i.e. the structure of the universe as a whole, including planets, comets, and stars), the earth (terrestrial phenomena, including meteorological matters), and finally the structure of the human body. All these topics are dealt with according to Cartesian principles, following the discussions of them found in Descartes's natural philosophical writings, chiefly the *Principles of Philosophy* and Descartes's *L'homme* ("Man"), which was originally written as a continuation of *Le monde* and finally published posthumously in 1662.

Rohault's basic tactic in all his exposition is to present the ideas and arguments as things that can be shown to carry inherent plausibility; he does not rely on invoking Descartes's name as an authority to support what he says (fittingly, given his remarks in the preface). For example, in a chapter on the three material elements making up the world, Rohault follows Descartes's tactic of imagining how matter might have taken on its present form naturally, regardless of God's true rôle in having shaped it as He willed. The relative motions of the parts of matter will have served, he says, to distinguish them from one another:

> This being supposed, it cannot be but that all these Particles of Matter must be broken where-ever [*sic*] they are angular, or are intangled with those that join to them; so that those which were supposed before to be very small, must become still smaller and smaller, till they are got into a Spherical Figure. Thus we have two Sorts of Matter determined, which we ought to account the two first Elements. And of these two we here call that which consists of the *very fine dust* which comes off from those Particles, which are not quite so small, when they are turned round, the *first Element*. And these particles thus made round, we call the *Second Element*. And because it may be, that some of the small Parts of Matter, either singly or united together, may continue in *irregular and confused Figures*, not so proper for Motion, we take them for the *third Element*, and join them to the other two.[9]

Rohault's entire argumentative strategy is one that, unlike Descartes's, does not aim at producing a systematically generated world-picture, but instead attempts to provide plausible, picturable, mechanistic explanations of phenomena on a rather piecemeal basis, sometimes drawing on analogies with technical processes. Like Huygens, Rohault promulgates a version of Cartesian physics that stresses a language for talking about physical matters intelligibly – which means *mechanistically*. A necessary internal consistency or interconnectedness is not a major desideratum, just as long as every individual explanation or exposition seems in itself to *make sense*.

Rohault died in 1672, but his place as a leading French exponent of Cartesian-style mechanistic natural philosophy was taken by a man called Pierre-Sylvain Régis, who came to Paris in 1680. Régis differed from Rohault, however, chiefly in his retention of an authentically Cartesian concern with *system*, involving a concern with the more metaphysical components of Descartes' philosophy. His great published work, the *Système de philosophie* ("System of Philosophy," 1690), covered logic, metaphysics, and moral philosophy as well as physics. It attempted, by including material such as some of Robert Boyle's experimental studies, to consolidate and promote an entirely Cartesian kind of philosophy that, in this respect, lacked the pragmatism and non-dogmatism of the mechanistic natural philosophy explored by Huygens and Rohault.

As a serious Cartesian philosopher, however, Régis had already been outpaced in France by another philosopher who, in the eyes of most of his contemporaries, was the pre-eminent exponent and developer of Descartes's philosophy in the final decades of the century. This was Nicolas Malebranche, a priest who was above all concerned with establishing a theologically orthodox version of Cartesianism, the religiously questionable character of which had proved troublesome on occasion for Rohault, among others. Malebranche's 1674/5 *Recherche de la vérité* ("Investigation of Truth") became an enormously important treatment of Cartesian philosophy, albeit one that had little to say about the concrete physical questions of especial concern to working natural philosophers like Huygens. As a result of such attempts, fully Cartesian in spirit, to infiltrate traditional university philosophy curricula by speaking to questions already established from Aristotelian texts, Cartesian ideas had begun to find their way into French university courses by the final decades of the seventeenth century. These ideas were not always cited with approval, but they were establishing themselves as genuine alternatives to older scholastic-Aristotelian approaches.

Cartesianism also found its champions in the fashionable salons of Paris. These upper-class retreats for intellectuals and would-be intellectuals grew in number and social significance in the second half of the seventeenth century (and on through the eighteenth), and became important shapers of opinion in the world beyond the universities and academies. In sharp distinction to those other forums for the discussion of ideas, the salons were presided over by women, and both men and women participated in their activities. The salons took the form of élite "open houses," generally occurring during designated afternoons each week, at the home of some gentleman of note who was, more usually than not, uninterested in the whole business. The proceedings would typically be overseen by the nobleman's wife, and great social prestige attached to the illustriousness of the literary and philosophical figures whom she could attract to the gatherings. Several women writers on philosophical as well as literary subjects in France during this period were participants in the gatherings of Madeleine de Scudéry (sister of a dramatist and member of the Académie Française, Georges de Scudéry) or of the Marquise de Rambouillet. Salon culture gave such women an opportunity, rare elsewhere, to discourse on an equal footing with men on matters generally restricted, as regards formal education, to a masculine clientèle. Apart from closeted private reading rather than active exchange and discussion, the only other significant formal opportunity for women in France to come into contact with philosophical, and specifically natural-philosophical, learning was at some of the public lectures given in Paris: Rohault explicitly opened his lectures to women in the mid-1660s. From this greatly circumscribed arena of opportunity and perceived relevance there emerged, however, several women whose philosophical views co-opted Cartesian arguments to justify their own place, as women, within philosophical conversation – in effect, underpinning their places in the new salon culture.

In brief, one of the inferences that could be drawn from Descartes's teachings, and was in fact drawn by a few men as well as women in this period, was this: because the mind is distinct from the body – being *res cogitans* rather than the ordinary matter of our corporeal frames – then there is in fact no fundamental difference between the minds of men and the minds of women: a mind is just a mind, with rationality as one of its hallmarks. As a consequence, women should be just as capable, intellectually, as men, and, with appropriate allowance being made for the universally accepted greater frailty of women's bodies, could participate in educational and even political pursuits like their male counterparts. Thus argued François Poullain de la Barre, a man whose talent for catchphrases yielded some popular success to his aphorism "the mind has no sex." Some women of the salons, including Catherine Descartes, the great philosopher's niece, contested such a stark mind–body dualism while at the same time engaging with it, as had the Princess Elizabeth, an important correspondent of Descartes himself in his later years. Such contested arguments act as demonstrations of the great flexibility of any philosophical system in relation to social and political questions.[10]

Within specifically natural-philosophical contexts, the kind of work done by Huygens left its mark, at the end of the seventeenth century and beginning of the eighteenth, in mathematical-physical work (that is, "physicomathematics") by members of the Swiss clan of mathematicians, the Bernoullis. Members of the Bernoulli family, from the 1690s through much of the eighteenth century, carried out pioneering theoretical work in fluid mechanics as part of a fundamentally Cartesian physical research programme that regarded all physical action as explicable in terms of matter pushing on matter, and in which motions in fluid media, such as Descartes's prototypical vortex motion, were of central theoretical importance. Sophisticated mathematical work on fluid mechanics around the middle of the eighteenth century was also conducted by another Swiss, Leonhard Euler, as part of an enterprise that was known in that century as "rational mechanics." But the prime mover in the establishment of eighteenth-century rational mechanics was not the work of Descartes or his direct followers, but the work of Isaac Newton.

II Newtonian alternatives

It should first be stressed that by the end of the seventeenth century there were *not* simply two monolithic, competing schools of thought labelled "Cartesian" and "Newtonian." As we have just seen, the work of people who regarded themselves as Cartesians was by no means uniform in its stress or detailed content, and the same general point also applies to "Newtonians." Before proceeding to Newton and his followers, however, notice must be taken of another important figure of the later seventeenth century who himself owed a lot to Descartes's example: the German philosopher Gottfried Wilhelm Leibniz.

Leibniz was an extraordinary figure who aspired to mastery of practically all fields of learning, from mathematics and logic to history and linguistics. From the 1670s until his death in 1716, Leibniz was in the service of the ruling family of the German state of Braunschweig, or Brunswick, and based in Hanover. His importance as a philosopher derives especially from his critiques of Descartes and his rejection of some of Descartes's most basic teachings. In natural philosophy, his most significant work occurred in theoretical physico-mathematical areas, especially mechanics, and in the latter he, rather than Newton, was responsible for some of the central concepts of the rational mechanics of the eighteenth century. Leibniz's natural philosophy was, even more than Descartes's, of a markedly metaphysical turn of mind. One of Descartes's most difficult and controversial doctrines had been the absolute distinction between mind (*res cogitans*) and body (ordinary matter/extension) in the human being, in which, nonetheless, the two were necessarily intimately connected. The philosophical difficulty that resulted from this position concerned how the mind and body could causally interact if they were utterly different from one another, the mind having no material, or mechanical, properties, and the material body having no mental properties. Descartes had never satisfactorily answered this question, as far as most subsequent philosophers were concerned, and Leibniz's solution was particularly radical. He proposed a "pre-established harmony" between mind and body, whereby God had arranged matters such that whatever the mind wills or experiences is exactly, but in fact causally-independently, matched by the physical goings-on of the material world. Thus, when I decide to kick a stone, and do so, and the stone moves (which I also observe to happen), all the physical components of that sequence occur utterly independently of the mental components. Leibniz's solution to Cartesian mind–body dualism was thus a kind of shadow-boxing.

Leibniz's quarrels with Newton were much more direct and vicious; not least, perhaps, because of the criticisms unleashed on Leibniz's character by Newton's cronies over priority in the invention of the calculus. Newton had first invented a form of the infinitesimal calculus in the 1660s, developing his first ideas on the subject in 1665–6, the same year that saw his original work on light and colours, discussed above in Chapter 7. That year is often called Newton's *annus mirabilis*, or "wonderful year," because, as well as those two signal inventions, he also began at that time the work on gravitation for which he is most famous.

By the mid-1660s Newton was well-versed in contemporary natural philosophy, including that of Descartes. His notebook from that period still survives, containing ideas, reflections, observations on reading, and experiments. The year 1665–6 was one that he spent in exile from the University of Cambridge (where he had just graduated BA) due to the presence in the city of the plague; Newton fled to his family estate in Grantham, Lincolnshire. Newton's ideas about gravity developed initially from his speculations on a problem posed by Galileo in his 1632 *Dialogo*, which had recently been translated from Italian into English. Galileo had considered why, if the earth spins on its axis, objects on its surface do not fly off, much as they will from a potter's wheel. This question prompted Newton to wonder what the centrifugal force at the earth's surface would be. (His considerations were independent of Huygens's still-unpublished work of the 1650s on centrifugal force, and its associated terminology.) Newton then also wanted to compare this outward-tending force with the force of gravity that nonetheless drew bodies inwards to the centre of the spinning earth.

The outcome was an analysis of motion in a circle that mirrored Huygens's (both being versions of the familiar formula $F = (mv^2)/r$. Newton used the motion of the moon around the earth as a check on this result, since he knew both the speed of the moon in its orbit and its approximate distance in terms of earth radii. If the moon behaved in the same kind of way as a body near the earth's surface, and its centrifugal tendency was exactly balanced by its gravitational tendency towards the earth, then Newton's formula implied that the gravitational force acting upon the moon had decreased from its measurable strength at the surface of the earth by a factor of $(1/r^2 - 1/R^2)$, where r is the earth's radius and R the moon's orbital radius. Newton claimed in after-years that he dropped the analysis at this point because he had used an erroneous figure for the earth's radius, which had thrown off the agreement between the inverse-square result and the observed behaviour of the moon. However that may be, Newton does not appear to have come back to these questions in any serious way until the late 1670s.

Edmund Halley's famous visit to the now-celebrated Lucasian professor of mathematics at Cambridge in 1684 was what prompted Newton to commence work on his great work *Philosophiae naturalis principia mathematica* ("The Mathematical Principles of Natural Philosophy"), published in 1687. Halley, acting as a sort of emissary of the Royal Society, wanted to know what resultant path would be traced out by a body in orbit around a stationary central body, if the moving body were attracted to the stationary one by a force that varied inversely as the square of their separation. The question assumed what we now know as "rectilinear inertia" for the motion of a body unaffected by any force acting on it from outside; that is, the body will continue to move in a straight line unless anything acts on it to deviate it from that course. This was a well-accepted principle by now, having been published by Gassendi in 1642 and, most notably, by Descartes in his 1644 *Principles of Philosophy*, and Newton too took it for granted. Newton answered Halley that the path would be an ellipse, just like the planets around the sun, and Halley encouraged Newton to publish the result. Doing so took Newton another two years' hard work, because he needed now to iron out the principles of an entire system of mechanics and motion, and coordinate it with experimental and observational data so as to apply it to the earth and solar system.

Not only Descartes's much-criticized rules of motion and collision, but Huygens's results on centrifugal force, first published without proofs in 1673, formed the background to Newton's work. Besides rectilinear inertia, the rules of collision that Huygens had already derived found their counterparts in Newton. What was of greatest importance for Newton in this work was to cast the entire treatment in classical geometrical form, to establish with solidity and mathematical rigour his comprehensive treatment of motion, intended to culminate in a "system of the world." His manuscript drafts of material found in the *Principia* are known, collectively, as "De motu" ("On Motion"), and date primarily from 1685. It was once thought that Newton must have derived his theorems using the calculus, only subsequently translating them into the terms of classical geometry to render them more acceptable to his contemporaries. All the evidence of the manuscripts, however, shows that he worked in the classical style from the beginning. Deductive, Euclidean-style geometry was still the appropriate language in which to perform such work, just as Latin was still the appropriate language in which to write mathematical texts.

One of the most notable features of the *Principia mathematica*, in contrast to Descartes's *Principia philosophiae*, is that it does not require its analyses to make use of direct contact between bodies as the means of transferring action. Newton speaks of "forces," prototypically understood in terms of discrete impulses, which act on a body so as to change its velocity (i.e. its speed or direction of motion, or both). He does not feel it incumbent upon him to provide a mechanism by which the force is communicated, or even, in all cases, to identify its source. Take, for example, the case of the problem

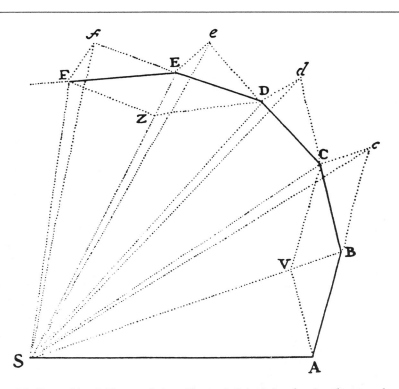

Figure 8.2 *Proposition I, Theorem I, from Newton's* Principia, *showing the general equal-area law. Note the impulse approach to centripetal force.*

brought to him by Halley, that of determining the path of a body orbiting a second, fixed body with an inverse-square law force acting between the two. Newton treated the path as if it consisted of successive rectilinear inertial motions punctuated by periodic discrete impulses towards the central body. This yielded a polygonal path which, when taken to the limit (a polygon with an infinite number of sides), yielded the curve that Newton sought. Newton disregarded the source or cause of the impulses in such analyses: in the *Principia*'s Proposition I, Theorem I, concerning what amounts to Kepler's second (equal-area) law, Newton simply identifies each of these discrete forces as "a centripetal force [that acts] at once with a single but great impulse."[11] Notice, however, that in leaving aside questions of causation, Newton appears to avoid the issue of whether any such force is an attractive force or a repulsive one. That is, the "great impulse" towards the central body might be some kind of attraction exerted *by* the central body, or it might be a *push* from outside *towards* the central body; Newton's analysis sidesteps any determination of the question.

However, Newton's original consideration in the mid-1660s of the moon's centrifugal force and the gravitational force acting so as to balance it clearly treated gravity as an *attraction* towards the earth. Newton's qualitative natural philosophy, as laid out in various places, but especially and most publicly in parts of his later book *Opticks*, makes it quite clear that he imagined gravitational attraction to be just that – a mutual attraction of one body for another whereby the attracted body is, so to speak, drawn in by the attractor. This is in contrast to the Descartes–Huygens way of understanding gravity, whereby heavy bodies are pushed towards the centre by the action of matter that is *further away* from that centre than they are.

Newton's complete "System of the World," which constitutes Book III of the *Principia*, shows with especial clarity the difficulties to which Newton's attempted finessing of the subject led him. Book III applies Newton's earlier mathematical investigations to the observed behaviour of bodies in the solar system, and shows how Kepler's laws of planetary motion could be derived from Newton's physico-mathematical assumptions once it was accepted that all material bodies attract one another as the inverse-squares of their distances (more precisely, that any two bodies attract one another with a force that varies as the inverse-square of the distance between their centres of gravity). But Newton *would not* specify what the cause of such a force was. Thus in the *Principia*:

> I use the word "attraction" here in a general sense for any endeavor whatever of bodies to approach one another. . . . I use the word "impulse" in the same general sense, considering in this treatise not the species of forces and their physical qualities but their quantities and mathematical proportions.[12]

And later, in his *Opticks*, where he discusses distance-forces in general, whether gravitational or not: "How these Attractions may be perform'd I do not here consider. What I call attraction may be perform'd by impulse, or by some other means unknown to me. I use that Word here to signify only in general any Force by which Bodies tend towards one another, whatsoever be the Cause."[13]

Part of Newton's difficulty was the mechanical philosophy itself. In one form or another – but always with at least implicit reference, whether approving or not, to Descartes's influential version – the mechanistic explanatory ideal hung over huge areas of non-Aristotelian, especially physico-mathematical, natural philosophy. Newton had absorbed its tenets and sensibilities as much as anyone else, and it was clearly difficult for him, in public at least, to deny them unequivocally. In 1692, in private correspondence on the subject, he wrote to Richard Bentley (later Master of Trinity College, Cambridge) as follows:

> It is inconceivable that inanimate brute matter should, without mediation of something else which is not material, operate upon and affect other matter without mutual contact. . . . Gravity must be caused by an agent acting constantly according to certain laws, but whether this agent be material or immaterial I have left to the consideration of my readers.[14]

Newton's need for something to mediate between attracting bodies might be considered the ghost of strict mechanism: if *not* matter as the mediator, then something else that plays the same rôle? Newton at one point played with the idea that God Himself brought about gravitational behaviours, by making bodies move according to the gravitational laws directly and without use of any intermediate physical cause whatsoever. That latter idea chimes nicely with Newton's famous remark, in the "General Scholium" to the *Principia*'s second edition (1713), concerning God: "He endures always and is present everywhere, and by existing always and everywhere he constitutes duration and space. . . . God necessarily exists, and by the same necessity he is *always* and *everywhere*. It follows that all of him is like himself: he is all eye, all ear, all brain, all arm, all force of sensing, of understanding, and of acting, but in a way not at all human, in a way not at all corporeal."[15] In Query 31 to the *Opticks*, he wrote that the evidently designful properties of animals, just like similar features of the solar system, "can be the effect of nothing else than the Wisdom and Skill of a powerful ever-living Agent, who being in all Places, is more able by his Will to move the Bodies within his boundless uniform sensorium."[16]

The natural-philosophical doctrine, or position, or ideology, known as "Newtonianism" largely revolved around these sorts of issues rather than around technical mathematical questions. Among Newton's earliest followers were such churchmen as Bentley, who promulgated versions of Newton's world-picture in order to promote particular theological and political goals. A prominent early forum for the promulgation of Newtonianism was an annual lectureship established by the will of Robert Boyle (who died in 1691), known, unremarkably, as the Boyle Lectureship – which still exists. Bentley was the first Boyle Lecturer, and he corresponded with Newton to get tips on how to use natural philosophy for supporting the Christian religion "against notorious infidels," as Boyle had put it. Newton was all in favour, telling Bentley: "When I wrote my treatise upon our Systeme [i.e. the *Principia*] I had an eye upon such Principles as might work with considering men for the beliefe of a Deity & nothing can rejoyce me more than to find it usefull for that purpose."[17]

Subsequent Boyle lecturers into the early eighteenth century, such as Samuel Clarke, William Whiston, and (particularly popular) William

Derham, also made use of versions of Newton's views on nature and on God's relationship to it. The importance of the various series of Boyle Lectures for the promulgation of Newtonianism lay in part in the fact that nearly every one of the sets of lectures was published in book form. Thus the first exposure to Newtonian natural-philosophical ideas for many members of the educated classes in Britain at the beginning of the eighteenth century came from their theological packaging by certain of the Boyle lecturers.

III Newtonianism

Newtonianism as an identifiable movement, with its characteristic philosophical style and its loyal adherents, really took shape after Newton became President of the Royal Society in 1703. (He had been living in London since his departure from Cambridge in the 1690s to take charge of the mint.) He ruled the Society until his death in 1727, and a sort of philosophical orthodoxy was the result. The establishment of this orthodoxy occurred to a significant extent through the actions of surrogates of Newton's in combating criticisms of Newton's work from other, usually Continental, philosophers. Of these critics, Leibniz stood first and foremost.

Leibniz, Huygens, and other Continental philosophers such as Régis had reacted critically when Newton published the *Principia* in 1687. Their main objections amounted to dismissing the pretensions of Newton's book: rather than presenting a true work on natural philosophy, Newton had simply presented mathematical description dressed up as natural philosophy. The author (perhaps Régis) of a review in the *Journal des Sçavans*, the leading philosophical journal in France, summed up his criticisms in this way: "In order to make an *opus* as perfect as possible, M. Newton has only to give us a Physics as exact as his Mechanics. He will give it when he substitutes true motions for those that he has supposed."[18] The criticism denies that a purely mathematical *description* ("Mechanics") can yield a physical *explanation*. The "true motions" sought by the reviewer would be ones the causes for which have been provided; Newton had merely "supposed" (that is, postulated) motions corresponding to gravitational forces without accounting for those forces in any way. Huygens's response was similar; we have already seen his explanation of gravity, developed originally well before the appearance of the *Principia,* and what he sought with his own account of gravity was what he failed to find in Newton's work – a physical explanation of gravitational behaviour.

Leibniz confronted relevant questions of planetary orbital motion in his *Tentamen de motuum coelestium causis* ("Essay on the Causes of Celestial Motions"). This monograph appeared in 1689 in the *Acta eruditorum,* a new, learned German review journal (by no means restricted to natural philosophy) based in Leipzig. Leibniz's essay was written partly in response to a review of the *Principia* in the *Acta*, and before, he implied, he had seen the *Principia* himself. While acknowledging the existence of gravitational attraction as apparently demonstrated by Newton, Leibniz was, like Huygens, concerned to *explain* it. He attempted to do so in terms of "lines of impulse" tending outwards from an attracting body through some kind of vortical fluid; this outward, centrifugal tendency brought about in turn reciprocal tendencies inwards by "terrestrial bodies" towards the centre – much as in the models of Huygens and Descartes.

Leibniz's subsequent battles with Newton occurred via proxies, and really came to a head two decades after this first, glancing encounter between the two. In the second edition of the *Principia*, in 1713, Newton's disciple, and the editor responsible for much of the work of producing the new edition, Roger Cotes, struck back at Cartesian-style critics of Newton's achievement. Cotes's preface ridicules the attempts of those, including Leibniz, who had postulated aethers and atmospheres of various kinds to account for the phenomena, accusing them of generating "an ingenious romance" on the basis of their (likely, false) conjectures. Explicitly naming, as enemies, those who follow "the opinions of *Descartes*," Cotes sternly asserts that:

> It is the province of true philosophy to derive the natures of things from causes that truly exist, and to seek those laws by which the supreme artificer willed to establish this most beautiful order of the world, not those laws by which he could have, had it so pleased him.[19]

Leibniz, in picking up this gauntlet not too long afterwards, therefore assailed the underlying metaphysical and theological assumptions of Newton's philosophy as a means of exposing its own erroneous "presuppositions" – Newton might want to hide behind a claim of causal nescience, restricting his arguments only to manifest and demonstrable facts, but Leibniz would show that Newton presupposed all manner of highly questionable positions concerning space, time, matter, and the Creator.

The ensuing debate took place through the medium of a published correspondence between Leibniz and Newton's surrogate, the former Boyle Lecturer Samuel Clarke. In the course of this exchange, which began in 1715 and was ended by the death of Leibniz in 1716, the most fundamental ideas in Newton's world-picture were placed under scrutiny. Leibniz characterized Newtonian gravitation as a "perpetual miracle," an unphilosophical concept that evaded the proper goals of philosophy. Newton's ideas of absolute space and absolute time were also, according to Leibniz, deeply flawed; Leibniz preferred relativity. Newton, egregiously, even regarded God as an imperfect clockmaker. This last brickbat was directed at Newton's belief, as expressed in the "General Scholium" to the *Principia*'s second edition, that the perfection of the solar system was compromised by the mutual gravitational attraction between the planets, which should disturb their regular orbits and ultimately throw the solar system into chaos. Newton liked this imperfection, because it allowed him to argue that God's active intervention was needed to prevent such a catastrophe: once in a while, God would fix the whole system up again before it got too much out of kilter, thereby evidencing his continuous presence in the universe and the existence of Divine Providence. For Leibniz, however, the necessity for such providence would be an imperfection in God, the clockmaker of the world mechanism.

The mustering of support for Newton's positions that occurred in the decades following Newton's ascent to the Presidency of the Royal Society was remarkably organized, in ways that are not always easy to explain. Newton carefully fostered his institutional power as the domineering president of the Society, the automatic deference shown him by most Fellows being translated into the encouragement and promotion of protégés who owed their own positions to Newton personally. This kind of personal patronage both within and without the Royal Society bred extraordinary, and deeply rooted, loyalty, as indicated in the rôles of Cotes and Clarke in combating critics of the *Principia*'s natural philosophy during the second decade of the eighteenth century.[20] Newton's work in optics, particularly as it appeared in his 1704 *Opticks*, met with similarly vigorous defences against foreign critiques of Newton's experimental inferences on the nature of light and colours. The Royal Society's official experimental demonstrators during Newton's presidency, holding the position of "Curator of Experiments", were, first, Francis Hauksbee, and later, from 1714, John Desaguliers (an Englishman of French Protestant – Huguenot – background). Both were loyal Newtonians who frequently used their experimental work to illustrate Newtonian ideas on such esoterica as the underlying nature of matter, and the attractive and repulsive forces that Newton conjectured, especially in later editions of the *Opticks*, to exist and operate at short ranges to produce such phenomena as electrical and chemical effects. Beyond the confines of the Royal Society, Desaguliers gave regular public lectures in London and published a widely read textbook, which went through several editions, called *A Course of Experimental Philosophy*. This printed version of his experimental demonstrations and philosophical teachings promoted to a much wider audience the Newtonian world-picture, based on distance-forces between particles, empty space, and the experimental foundations of natural philosophy. Like Newton, Desaguliers continued to contrast that picture with the Cartesian, with its contact-action transmission of forces between bodies, its universal space/matter, and its (supposedly) rationalist approach to understanding the universe.

In the eighteenth century, the spread of Newtonianism in England and, increasingly, in continental Europe, accompanied its association with the philosophy of John Locke, as laid out in Locke's *Essay Concerning Human Understanding* (1690). Locke had made it his business to investigate the proper foundations of knowledge, and to act, as he put it, as an "under-labourer" to the work of the Royal Society's experimentalists (he was a personal acquaintance of Boyle and Newton, and even assisted the former in some of his publications). The chief confluence of Locke's philosophy and the natural philosophy of Newton lay in Locke's stress on empiricism as the route to knowledge. Newton himself always characterized his discoveries as being founded on experiment and observation rather than on "innate ideas" of the sort chased by Descartes (think of the role of *cogito ergo sum*). Locke was generally read in the eighteenth century as having supported the same view of the sources of knowledge, constructed with much more elaborate arguments.

Two major areas of Newton's own thought and work, however, were largely purged from the Newtonianism of the eighteenth century: theology, including Newton's studies of biblical chronology and interest in biblical prophecy, and alchemy, which occupied him for many years, especially in the 1670s (see Chapter 1, section IV, above, for more on alchemy). Newtonianism after Newton stressed the kind of rational empiricism found in Newton's publications and in the elaborations of his followers – including particular kinds of theological inferences favourable to the new Anglican orthodoxy that had followed the Glorious Revolution of 1688, a political event that Newton himself, as a member of parliament for the University of Cambridge, had wholeheartedly supported.

The story of the continuing debates between "Cartesians" and "Newtonians" in the eighteenth century carries us well beyond the confines of this book. But it is worth observing that the story was not to be a simple one of Newtonian "truth" beating out Cartesian "romance" (as some critics liked to characterize Descartes's mechanical universe). The complexity and interweaving of arguments, mathematical, metaphysical and experimental, meant that even when, in the later decades of the eighteenth century, Newton's name was generally invoked as the winner of the supposed contest, what counted as "Newtonianism" was in many ways quite different from what Newton himself had believed and argued. The "Newtonianism" of the later eighteenth century was itself a hybrid of Newton's, Descartes's, Leibniz's, and many other people's work and ideas.

Conclusion
What was Worth Knowing by the Eighteenth Century?

By the time of Newton's death, the educated European outlook on the natural world had changed beyond all recognition from what it had been in 1500. The new ideology of natural knowledge was now one firmly, though not exclusively, associated with practical, operational capabilities. The greatest physico-mathematicians of the later seventeenth century, Huygens and Newton, both took an active interest in practical, non-contemplative matters. Significantly, in the 1650s, Huygens had devoted much attention to the problem of the determination of longitude at sea, a problem of especial concern to the new mercantile states of Western Europe such as Huygens's own nation of the Netherlands (United Provinces). In addressing it, Huygens not only dealt with the theoretical problems relating to pendulum motion (the use of the pendulum as a timekeeper had earlier been suggested by Galileo), but also worked on the details of actually constructing a marine chronometer that would continue to keep regular time on ocean voyages: Huygens's chronometers were actually put to practical trial on long voyages by French naval vessels. The incessant rhetoric of Baconian practicality that dominates the first decades of the Royal Society was also important for Huygens and the Royal Academy of Sciences in Paris, and it remained crucial in the early decades of the eighteenth century with the establishment of a Newtonian natural-philosophical ideology.

The major development of the two centuries covered in this book was, therefore, the rise to a position of prominence of a "natural philosophy" that was directed towards control of the world. European knowledge in 1500, as it existed in formal, official settings such as universities, placed a premium on abstract, contemplative understanding. This is not to say that there were no social implications of such a focus, but it is to say that those implications were mediated through institutions (especially the Church) whose power did not noticeably involve ambitions to increase the means of control over the natural world itself. During the sixteenth and seventeenth centuries, however, European nations began to spread their power to other parts of the world to an extent unprecedented in history. Consequently, valuations of knowledge began very gradually to shift towards those kinds of knowledge that could bring the world beyond Europe back home (as with geography and natural history), or that would enable a more effective reaching out to other parts of the world with the intention of material and cultural domination (as with such sciences as navigation or mechanics – or even with Matteo Ricci's use of mathematics to impress the Chinese court). The rise of a Baconian rhetoric of utility during the seventeenth century, associated with the welfare of the state, mirrored closely these large-scale changes in European life.

Significantly, it was the mercantile states of western Europe that played the greatest role in revolutionizing the sciences during this period. Spain, the greatest colonial power of the period in terms of wealth acquired and land conquered, but not the greatest as an active mercantile power, did not follow the same direction as countries such as France, England, or the Netherlands, except perhaps in studying the natural history of the New World. England and the Netherlands in particular illustrate well the associations between mercantile colonial expansion and the new ambitions of European knowledge in these centuries.

Concomitantly, while the sixteenth century had witnessed a form of intellectual endeavour that was dominated by humanism, and by the explicit aim of recovering the civilization of classical antiquity, the seventeenth century saw the appearance of a new ambition, exemplified by Descartes and Bacon, to forge ahead with professedly novel intellectual programmes. The sanction of antiquity remained an important rhetorical resource for many, but it now competed with claims of novelty that often justified approaches to nature by talk of "method" instead of talk about classical precedent. The evidence that such methods were efficacious was argued to reside in the practical achievements that the method supposedly enabled, whether it was Bacon's inductive method leading to "works," or Descartes's method leading to improved optical lenses (as in his essay "Dioptrics") or, as Descartes also hoped, to lengthened human lives.

All the same, the category of endeavour known as "natural philosophy" retained certain fundamental features right through all the changes that occurred during this period. From beginning to end, natural philosophy involved God, whether Thomas Aquinas's medieval God of an Aristotelian universe or the God of the Newtonians, free to do whatever He wanted and continually, providentially aware of everything in the universe due to His omnipresence throughout all of (absolute) space – what Newton called God's "universal sensorium." Natural philosophy bred very few genuine atheists in the sixteenth or seventeenth centuries, although matters changed in the eighteenth.

It would be foolish to see the so-called Scientific Revolution as nothing but a straightforward product of European expansion. The emergence in the seventeenth century of the infinite universes of Descartes and of Newton, with the earth a planet orbiting a star called the sun, can stand for enormous intellectual shifts in the kind of universe that educated Europeans saw themselves as inhabiting. Nonetheless, at the heart of these shifts are the operational, mathematical, and (in the case of natural history) enumerative or cataloguing enterprises of the period, enterprises that underpinned the creation of a new universe and a new natural philosophy. European learned culture, in regard at least to an understanding of the natural world, had undergone a shift from a stress on the *vita contemplativa*, the "contemplative life," to a stress on the *vita activa*, the "active life," to use a Latin terminology familiar to the humanist scholars of the period.[1] "Knowing *how*" was now starting to become as important as "knowing *why*." In the course of time, those two things would become ever more similar, as Europe learned more about the world in order to command it. The modern world is much like the world envisaged by Francis Bacon.

Notes and References

Introduction

1 See, e.g., Jean d'Alembert, *Preliminary Discourse to the Encyclopedia of Diderot* [1751], trans. Richard N. Schwab (Indianapolis: Bobbs-Merrill, 1963), p.80.
2 See below, Chapter 3, section IV.
3 Aristotle, *Posterior Analytics* II.19, in Jonathan Barnes (ed.), *The Complete Works of Aristotle*, 2 vols (Princeton: Princeton University Press, 1984), vol.1, pp.165–6 (trans. Barnes).
4 It is in the nature of heavy bodies to fall downwards. This in turn is understood in terms of "final causes": their natural place (i.e., their destination) is the centre of the universe, towards which they therefore tend to go. See Chapter 1, section I, below.
5 Francis Bacon, *The New Organon*, ed. and trans. Lisa Jardine and Michael Silverthorne (Cambridge: Cambridge University Press, 2000), Book I, aph.93. Bacon refers here to the voyages of discovery, using the image of passing through the Straits of Gibraltar into the great ocean beyond.
6 An insightful recent discussion of this very point is Peter Pesic, "Wrestling with Proteus: Francis Bacon and the 'Torture' of Nature," *Isis* 90 (1999), pp.81–94.
7 Hence the title of Marie Boas Hall's book *The Scientific Renaissance, 1450–1630* (New York: Harper & Row, 1962).
8 Nonetheless, important exceptions still remained, including features of the work of Isaac Newton: see J. E. McGuire and P. M. Rattansi, "Newton and the 'Pipes of Pan'," *Notes and Records of the Royal Society of London* 21 (1966), pp.108–43.

Chapter 1: "What was Worth Knowing" in 1500

1 Isaac Newton, *The Principia: Mathematical Principles of Natural Philosophy. A New Translation and Guide*, trans. I. Bernard Cohen and Anne Whitman (Berkeley, etc.: University of California Press, 1999), p.943.
2 Generally, the desideratum was to locate planets within the correct zodiacal sign, i.e. to within fifteen degrees, for astrological purposes.
3 Theology and philosophy during the Middle Ages themselves exhibit many "Platonic" strains, but there was little close study of Plato's own texts. Medieval Platonism had typically been refracted by other authors, especially the Christian Church Father St. Augustine (c.400 AD).

4   William Eamon, *Science and the Secrets of Nature: Books of Secrets in Medieval and Early Modern Culture* (Princeton: Princeton University Press, 1994), p.113.
5   On the importance of "tacit knowledge" in science, see H. M. Collins, *Changing Order: Replication and Induction in Scientific Practice*, 2nd edn (Chicago: University of Chicago Press, 1992).
6   The astronomer Tycho Brahe, in the late sixteenth century, was also an alchemical practitioner.
7   Eamon, *Science and the Secrets of Nature*, pp. 114–15.
8   The classic exposition of this view is Keith Thomas, *Religion and the Decline of Magic: Studies in Popular Beliefs in Sixteenth- and Seventeenth-Century England* (London: Penguin Books, 1978).

Chapter 2: Humanism and Ancient Wisdom: How to Learn Things in the Sixteenth Century

1   Separate humanist schools were also set up, seldom but sometimes including such education for girls.
2   Regiomontanus completed the *Epitome* after Peurbach's death in 1461. See also Chapter 1, section III, above.
3   Nicholas Copernicus, *On the Revolutions*, translation and commentary by Edward Rosen (Baltimore: The Johns Hopkins University Press, 1992), p.4.
4   Ibid., p.5.
5   Quoted in Alexandre Koyré, *The Astronomical Revolution: Copernicus – Kepler – Borelli* (London: Methuen, 1973), p.29.
6   See Chapter 1, section III, above.
7   Translated in C. D. O'Malley, *Andreas Vesalius of Brussels 1514–1564* (Berkeley, etc.: University of California Press, 1964), p.320.
8   Ibid.
9   Galen was particularly concerned to give explanations for anatomical features in terms of final causes.
10 Vesalius, in O'Malley, *Andreas Vesalius*, p.322.
11 Ibid.
12 "Coss" from the Italian "cosa," meaning "thing" – what we would call the "unknown" in an algebraic equation.
13 René Descartes, *The Philosophical Writings*, vol.1, ed. John Cottingham, Robert Stoothoff and Dugald Murdoch (Cambridge: Cambridge University Press, 1985), p.17 (trans. Murdoch).
14 That is, with a *stationary* sun, but not strictly helio*centric*, with the sun at the *centre* of the system – as Kepler later noted. In Copernicus's complete system, the sun was slightly off to one side of the centre of the earth's orbit, which provided the actual fixed point with which the other planetary motions were coordinated.
15 See Chapter 1, section III, above.
16 Copernicus, *De revolutionibus*, trans. Rosen, p.xx.
17 Ibid.
18 As long as the stars are now taken as being enormously distant compared to the size of the earth's orbit.
19 Guidobaldo's most important work was his *Liber mechanicorum* ("Book of Mechanics") of 1577.
20 Quoted in Paul Lawrence Rose, *The Italian Renaissance of Mathematics: Studies on Humanists and Mathematicians from Petrarch to Galileo* (Geneva: Droz, 1975), p.230.

Chapter 3: The Scholar and the Craftsman:
Paracelsus, Gilbert, Bacon

1 These are the Latin forms of their names, by which they were known in the universities.

2 Blood was hot and wet, phlegm was cold and wet, black bile was cold and dry, and yellow bile was hot and dry. The Aristotelian elements were also associated with these pairs of qualities.

3 Quoted in Walter Pagel, *Paracelsus: An Introduction to Philosophical Medicine in the Era of the Renaissance*, 2nd rev. edn (Basel and New York: Karger, 1982), p.71.

4 Ibid., p.84.

5 And see above, Chapter 1, section IV.

6 Cf. Chapter 2, section IV, above, on Pliny and Lutheran pedagogy. Agricola's book is also the chief literary source, through his presentation of traditional miners' lore, of the association between mining and dwarfs, the gnome-like beings that were believed to lurk in the mines: Georgius Agricola, *De re metallica*, trans. Herbert Clark Hoover and Lou Henry Hoover (New York: Dover, 1950), p.217 n.26.

7 See also Chapter 4, section IV, below.

8 William Gilbert, *De magnete*, trans. P. Fleury Mottelay (New York: Dover, 1958), pp.14–15.

9 Ibid., p.15.

10 Ibid., p.47.

11 Ibid., p.22.

12 See, e.g., Gilbert's self-presentation in ibid., "Preface," pp.xlvii-li.

13 For example, Francis Bacon, *The New Organon*, ed. and trans. Lisa Jardine and Michael Silverthorne (Cambridge: Cambridge University Press, 2000), Book I, aph.63, comparing Aristotle unfavourably to the Presocratics.

14 Ibid., aph.32.

15 Ibid.

16 Ibid., aph.122.

17 Ibid., aph.79.

18 Ibid., aph.71.

19 Ibid., aph.79.

20 Ibid., aph.33.

21 Ibid., aph.124.

22 Note the Aristotelian concepts here.

23 Bacon, *New Organon*, Book I, aph.66 (end); my emphasis.

24 Ibid., aph.81.

25 Ibid., aph.129 (end).

26 On the syllogism, see Introduction, section II, above.

27 Bacon, *New Organon*, Book I, aph.11.

28 Ibid., aph.8.

29 Ibid., aph.12.

30 Ibid., aph.105.

31 Ibid., aph.106.

32 For example, ibid., aph.100.

33 Ibid., aph.101. Bacon's term for this is, in Latin, *experientia literata*: "literate experience."

34 For more on Harvey, see below, Chapter 7, section IV.

35 Francis Bacon, *The Works*, 7 vols, edited by James Spedding, R. L. Ellis and D. D. Heath (London, 1857–61), vol.3, p.156.

36 Bacon, *New Organon*, Book I, aph.96.

37   Ibid., Book II, aphs.6, 7. "Spiritual" in that he believed – along with the alchemists – that some bodies, especially metals, had properties conferred on them by a subtle spirit that occupied the space between their particles.
38   Ibid., aph.5.
39   See his worked example of a determination of the true nature ("form") of heat, ibid., aphs.11–20.

Chapter 4: Mathematics Challenges Philosophy: Galileo, Kepler, and the Surveyors

1   See Chapter 1, section I, above.
2   See Chapter 1, section II, above.
3   Quoted in Peter Dear, *Discipline and Experience: The Mathematical Way in the Scientific Revolution* (Chicago: University of Chicago Press, 1995), p.35.
4   On Guidobaldo, see Chapter 2, section V, above.
5   See Chapter 2, section V, above.
6   Cf. section IV of the present chapter.
7   The preferred English translation of this work is Galileo Galilei, *Discourses and Demonstrations Concerning Two New Sciences*, trans. Stillman Drake (Madison: University of Wisconsin Press, 1974). For more on Galileo's career, see Chapter 6, section II, and Chapter 7, section I, below.
8   Michael Sharratt, *Galileo: Decisive Innovator* (Cambridge: Cambridge University Press, 1994), p.70; passage from Galileo trans. in Annibale Fantoli, *Galileo: For Copernicanism and for the Church*, trans. George V. Coyne (2nd edn, Rome: Vatican Observatory, 1996), p.70.
9   See Chapter 6, section II, below.
10   Copernicus had attempted to deal with this in Book I of *De revolutionibus* by postulating gravity (a tendency to coherence) as a property of each individual spherical celestial body, including the earth.
11   See above, Chapter 2, section IV.
12   See discussion in William R. Shea, *Galileo's Intellectual Revolution* (London: Macmillan, 1972), pp.55–7.
13   Translated by Stillman Drake in Drake (ed.), *Discoveries and Opinions of Galileo* (Garden City, NY: Doubleday Anchor Books, 1957), p.124.
14   Ibid.
15   See Dear, *Discipline and Experience*, p.103.
16   Translated by Nicholas Jardine in Jardine, *The Birth of History and Philosophy of Science: Kepler's "A Defence of Tycho Against Ursus" with Essays on its Provenance and Significance* (Cambridge: Cambridge University Press, 1984), p.250.
17   Witelo was a Polish writer on optics of the late thirteenth century.
18   Johannes Kepler, *Ad Vitellionem paralipomena* (Frankfurt, 1604; facs. reprint Brussels: "Culture et Civilisation," 1968), dedication, pp.2v–3r.
19   See esp. John Dee, *The Mathematicall Preface to the Elements of Geometrie of Euclide of Megara*, intro. Allen G. Debus (1570; facs. reprint New York: Science History Publications, 1975), sixth p. (unpaginated).

Chapter 5: Mechanism: Descartes Builds a Universe

1   Francis Bacon, *The New Organon*, ed. and trans. Lisa Jardine and Michael Silverthorne (Cambridge: Cambridge University Press, 2000), Book I, aph.66 (end).
2   See Descartes's own account in the *Discourse on the Method*, Pt.I, in John Cottingham,

Robert Stoothoff and Dugald Murdoch (trans. and ed.), *The Philosophical Writings of Descartes*, 3 vols (Cambridge: Cambridge University Press, 1985–91), vol.1, pp.115–16.
3   See Chapter 4, section III, above.
4   Isaac Beeckman, *Journal tenu par Isaac Beeckman de 1604 à 1634*, 4 vols, ed. C. de Waard (The Hague: Martinus Nijhoff, 1939–1953), p.206. See Stephen Gaukroger, *Descartes: An Intellectual Biography* (Oxford: Clarendon Press, 1995), chapter 3; note especially long quote on p.71 from John Schuster on Beeckman's artisanal criterion of intelligibility.
5   The same publisher also printed the first edition of Galileo's *Two New Sciences* in the following year, 1638 – Dutch publishing laws were notoriously liberal, for the period.
6   See, on the notion of "atoms of evidence," Marjorie Grene, *Descartes* (Minneapolis: University of Minnesota Press, 1985), p.54.
7   This was, in fact, a standard scholastic philosophical maxim, so Descartes could reasonably expect people to grant it to him.
8   All this in the *Discourse*, Pt.4; quote in Cottingham *et al.*, *Philosophical Writings of Descartes*, vol.1, p.127.
9   *Le monde, ou le traité de la lumiere* (*sic*) appeared in print in 1664, fourteen years after Descartes's death; his reasons for suppressing it in 1633 were that, as he wrote in a letter at the time, he had just heard of Galileo's condemnation in Rome for teaching the motion of the earth, which was an integral feature of Descartes's world-picture too. The work has recently been published in a new English translation: René Descartes, *The World and Other Writings*, trans. Stephen Gaukroger (Cambridge: Cambridge University Press, 1998).
10 Trans. in Cottingham *et al.*, *Philosophical Writings of Descartes*, vol.1, p.81.
11 *Principia philosophiae*, 1644, composed in Latin; *Principes de la philosophie*, a French translation authorized by Descartes, 1647.
12 *Le monde*, chapter 6 and ff. See also the similar statement in the *Principia philosophiae*, Pt.III, para.45 (Cottingham *et al.*, *Philosophical Writings of Descartes*, vol.1, p.256).
13 See Chapter 4, section III, above.
14 *Principia philosophiae*, Pt.II, para. 64.; trans. in Cottingham, *Philosophical Writings of Descartes*, vol.1, p.247.
15 "Common notions" refers, in Euclid, to the basic, universally accepted starting principles from which deductive demonstrations in mathematics are derived – supposedly intuitively obvious statements such as "When equals are subtracted from equals, the remainders are equal."
16 *Principia philosophiae*, Pt.II, para. 64.; trans. in Cottingham, *Philosophical Writings of Descartes*, vol.1, p.247.
17 Descartes gave this account in the "Dioptrics," one of the essays attached to the *Discourse*.
18 A good edition in English of Descartes's book is René Descartes, *The Passions of the Soul*, ed. and trans. by Stephen Voss (Indianapolis: Hackett, 1989). Descartes's major physiological writings are his *Treatise on Man* and *Description of the Human Body*, both translated in Descartes, *The World*, trans. Gaukroger.
19 The law has since become known as "Snell's law," after the Dutchman Willebrord Snell, who had derived, but did not publish, the relationship in 1621.
20 See *Discourse*, end of Part VI (Cottingham, *Philosophical Writings of Descartes*, vol.1, pp.150–1).
21 Ibid. (trans. Cottingham, *Philosophical Writings of Descartes*, vol.1, p.149).
22 For example, Plate XIII in *Principia philosophiae* gives a representation of the point that was made with the example of the wine vat in the "Dioptrics," regarding simultaneous tendencies in many directions at once. This time, however, rather than using an analogy with wine, Descartes uses representations of what are purported to be the physically real material bodies that actually communicate light.
23 Trans. Cottingham, *Philosophical Writings of Descartes*, vol.1, pp.86–7.

24 Trans. Cottingham, *Philosophical Writings of Descartes*, vol.1, p.241; see plate in ibid., p. 259.
25 Trans. Miller and Miller, in René Descartes, *Principles of Philosophy*, trans. Valentine Rodger Miller and Reese P. Miller (Dordrecht: D. Reidel, 1983), p.61.
26 Ibid., pp.108–10, and Descartes, *The World*, chapter 5. Thus fire, for example, is also a manifestation of first element in agitation. In addition, because the globules of second element in the heavens are of a fixed, roughly spherical shape, and there is no such thing as empty space, the interstices between them are necessarily also filled with matter – yet more particles of fluid first element which can accommodate themselves to the shape of any gap. These latter particles of first element do not typically bring forth light because they are not sufficiently agitated.
27 This was the perspective of John William Draper, *History of the Conflict Between Religion and Science* (New York: 1874), and Andrew Dickson White, *A History of the Warfare of Science with Theology in Christendom* (London: 1896).
28 *Principia philosophiae*, Pt.2, para.25; trans. Cottingham, *Philosophical Writings of Descartes*, vol.1, p.233.
29 Ibid., Pt.3, para.26, trans. Miller and Miller, *Principles*, p.94. The words in brackets represent authorially approved interpolations found in the French translation of 1647.
30 But his matter theory and the Eucharist caused trouble later on: see Richard A. Watson, "Transubstantiation among the Cartesians," in Thomas M. Lennon, John M. Nicholas and John W. Davis (eds), *Problems of Cartesianism* (Kingston and Montreal: McGill-Queen's University Press, 1982), pp.127–48, together with the criticisms in Roger Ariew, *Descartes and the Last Scholastics* (Ithaca: Cornell University Press, 1999), esp. pp.141–2.
31 Descartes discusses magnets and magnetism at some length in the *Principles*: pp.242–72 in Miller and Miller translation.
32 René Descartes, "Meteorology," in Descartes, *Discourse on Method, Optics, Geometry, and Meteorology*, trans. Paul J. Olscamp (Indianapolis: Bobbs-Merrill, 1965), pp.263–361, on pp.275–6.

Chapter 6: Extra-Curricular Activities: New Homes for Natural Knowledge

1 Cf. Chapter 1, section III, above.
2 Quoted in Thomas S. Kuhn, *The Copernican Revolution: Planetary Astronomy in the Development of Western Thought* (Cambridge, Mass.: Harvard University Press, 1957), p.191.
3 Nicholas Copernicus, *On the Revolutions*, trans. Edward Rosen (Baltimore: Johns Hopkins University Press, 1992), p.XX.
4 Robert S. Westman, "The Astronomer's Role in the Sixteenth Century: A Preliminary Study," *History of Science* 18 (1980), pp.105–47, esp. p.107.
5 See above, Chapter 2, section II.
6 Quoted in William R. Shea, *Galileo's Intellectual Revolution* (London: Macmillan, 1972), p.14.
7 Original title page reproduced in Johannes Kepler, *New Astronomy*, trans. William H. Donahue (Cambridge: Cambridge University Press, 1992), p.26; and see Figure 6.1.
8 Evidently intended to mean "starry (or sidereal) message," although it was taken by many, including Kepler, to mean "starry messenger," an equally proper reading of the Latin and the one by which it is usually known in English.
9 Galileo Galilei, *Sidereus nuncius, or, The Sidereal Messenger*, trans. Albert Van Helden (Chicago: University of Chicago Press, 1989), p.62.

10 See also Chapter 4, section II, above, on this earlier mechanical work, as well as Chapter 7, section I, below.

11 Biagioli notes this difference in his "Scientific Revolution, Social Bricolage, and Etiquette," in Roy Porter and Mikuláš Teich, *The Scientific Revolution in National Context* (Cambridge: Cambridge University Press, 1992), pp.11–54, on p.51 n.105. Biagioli speaks of Harriot's failure to seek a "courtly representation" of his discoveries along the lines later to be travelled by Galileo as a reflection of the greater diversity of less rigidly hierarchized venues for philosophical activity in England. It may also reflect the fact that Harriot already had a stable patronage relationship that suited him and that established him reasonably independently outside the universities.

12 See Chapter 7, section IV, below for more details.

13 Translation modified from William Harvey, *The Circulation of the Blood and Other Writings*, trans. Kenneth J. Franklin (London: Dent, 1963), p.3.

14 Quoted in Walter Pagel, *William Harvey's Biological Ideas: Selected Aspects and Historical Background* (New York: Hafner, 1967), p.19 (from the *Letters to Riolan*, 1649).

15 Trans. Franklin in Harvey, *Circulation*, p.161.

16 Ibid., p.6.

17 See Paolo Rossi, *Francis Bacon: From Magic to Science*, trans. Sacha Rabinovitch (Chicago: University of Chicago Press, 1968), p.153 (discussing Bacon's *Advancement of Learning*), and Chapter 4, section IV, above.

18 See Chapter 3, section III, above.

19 Besides Bacon, there were many projectors of encyclopedic attempts to encompass all knowledge, such as Jean Bodin: see Ann Blair, *The Theater of Nature: Jean Bodin and Renaissance Science* (Princeton: Princeton University Press, 1997). On the *Accademia* and natural history, see Paula Findlen, *Possessing Nature: Museums, Collecting, and Scientific Culture in Early Modern Italy* (Berkeley, etc.: University of California Press, 1994), pp.31–3.

20 Trans. Franklin in Harvey, *Circulation*, p.3.

21 See Chapter 8, section I, below.

22 Quoted in Roger Hahn, *The Anatomy of a Scientific Institution: The Paris Academy of Sciences, 1666–1803* (Berkeley, etc.: University of California Press, 1971), p.25.

23 They did, however, receive the not inconsiderable right to approve books for publication, a right that the government normally held to itself. See Michael Hunter, *Science and Society in Restoration England* (Cambridge: Cambridge University Press, 1981), p.36, on their cautious use of this privilege.

24 Thomas Sprat, *History of the Royal Society* (London, 1667; facsimile reprint, Saint Louis: Washington University Press, 1958), p.53.

25 There are recent editions of works by Conway and Cavendish, including Anne Conway, *The Principles of the Most Ancient and Modern Philosophy*, ed. Allison Coudert and Taylor Corse (Cambridge: Cambridge University Press, 1996), which illustrates the theological importance of natural philosophy; Margaret Cavendish, *Grounds of Natural Philosophy*, intro. by Colette V. Michael (West Cornwall, Conn.: Locust Hill Press, 1996); Cavendish, *Paper Bodies: A Margaret Cavendish Reader*, ed. Sylvia Bowerbank and Sara Mendelson (Peterborough, Ontario: Broadview Press, 2000); Cavendish, *The Description of a New World Called the Blazing World and Other Writings*, ed. Kate Lilley (London: Pickering, 1992), this latter volume containing imaginative and moral writings. See also Chapter 8, section I, for more on women as participants in the culture of natural philosophy.

26 Quoted in Svetlana Alpers, *The Art of Describing: Dutch Art in the Seventeenth Century* (Chicago: University of Chicago Press, 1983), p.17.

27 See above, Chapter 3, section III.

Chapter 7: Experiment: How to Learn Things about Nature in the Seventeenth Century

1 See Paul Cranefield, "On the Origins of the Phrase *Nihil est in intellectu quod non prius fuerit in sensu*," *Journal of the History of Medicine*, 25 (1970), 77–80.

2 Francis Bacon, *The New Organon*, ed. and trans. Lisa Jardine and Michael Silverthorne (Cambridge: Cambridge University Press, 2000), Book I, aph.63.

3 Galileo Galilei, *Dialogue Concerning the Two Chief World Systems*, trans. Stillman Drake (Berkeley, etc.: Univ. of California Press, 1967), p.234.

4 Cf. Chapter 6, section V, above, on "discovery" in the seventeenth century.

5 See Chapter 1, section I, above.

6 See Chapter 4, section II, above, on the work's earliest stages.

7 All such statements, whether natural or mathematical, were of course always open to criticism from philosophical sceptics; see Chapter 5, section I, above.

8 René Descartes, *The Philosophical Writings of Descartes*, trans. John Cottingham, Robert Stoothoff and Dugald Murdoch, 3 vols (Cambridge: Cambridge University Press, 1985–91), vol.1, p.148.

9 Chap.4, section III.

10 Alhazen was known in Arabic as Ibn al-Haytham.

11 Calling the apparatus a "mercury barometer" thus begs the whole question, since the entire test was aimed at showing that this apparatus *was* a "barometer," a "measurer of weight," meaning weight, or pressure, of the ambient air.

12 Steven Shapin, *A Social History of Truth: Civility and Science in Seventeenth-Century England* (Chicago: University of Chicago Press, 1994), pp.266–91, presents an interesting dispute revolving around these issues.

13 See Chapter 3, section III, above.

14 Bacon, *New Organon*, Book II, aph.11.

15 Ibid., aph.13, item 28.

16 Robert Boyle, *An Hydrostatical Discourse*, in Robert Boyle, *The Works of the Honourable Robert Boyle*, ed. Thomas Birch, 6 vols (London, 1772; facsimile reprint, Hildesheim: Georg Olms, 1965–66), vol.3, p.611.

17 Ibid., p.612. The experiment involved a cylinder of wax that would sink or float in water depending on very small weight changes occasioned by the addition or subtraction of small pieces of brass – the point being that the density of the wax itself was only very slightly less than that of the water.

18 Robert Hooke, *Micrographia, or Some Physiological Descriptions of Minute Bodies Made by Magnifying Glasses* (London: 1665), "To the Royal Society."

19 Thomas Hobbes, *The English Works of Thomas Hobbes*, ed. Sir William Molesworth, 11 vols (London: 1839–1845), vol.7, p.184 (from *Six Lessons to the Mathematicians*).

20 Boyle, *Works*, vol.3, p.610.

21 It is also significant that Hobbes never challenged the truth of Boyle's experimental assertions, but only challenged Boyle's causal understanding of the phenomena. In this regard, Hobbes's objection to the "experimental philosophy" was not that it yielded false results, but simply that it was not philosophy.

22 From the *Saggi*, translation in W. E. Knowles Middleton, *The Experimenters: A Study of the Accademia del Cimento* (Baltimore: The Johns Hopkins University Press, 1971), pp.246–7. The physical presuppositions that motivate this experiment, whereby qualitative physical phenomena are automatically conjectured as being manifestations of minute corpuscles of some kind, is another point of similarity between the Cimento and the Royal Society. Despite many points of difference, they, and many other contemporary natural philosophers, apparently found imaginary particles to yield an especially intelligible *kind* of explanation. The specific idea here, of property-bearing particles of heat and cold,

appears to have been borrowed from Gassendi, although Galileo too spoke of heating in terms of the introduction of particles of fire into the warmed body.

23 See the reprint in Marie Boas Hall, *Nature and Nature's Laws: Documents of the Scientific Revolution* (New York: Walker and Company, 1970), p.250.
24 Ibid.
25 Ibid., p.251.
26 Ibid., p.255.
27 See Chapter 6, section III, above.
28 William Harvey, *The Circulation of the Blood and Other Writings*, trans. Kenneth J. Franklin (London: Dent, 1963), p.5.
29 Ibid., p.166.
30 Ibid., p.167.

Chapter 8: Cartesians and Newtonians

1 See Peter Dear, *Mersenne and the Learning of the Schools* (Ithaca: Cornell University Press, 1988), pp.210–11.
2 Christiaan Huygens, *Oeuvres complètes de Christiaan Huygens*, 22 vols (The Hague: Nijhoff, 1888–1950), vol.21, p.446 (my translation). This text was first published in 1690, in the wake of Newton's *Principia*.
3 Ibid., vol.19, p.631.
4 See Chapter 5, section III, above.
5 Huygens, *Oeuvres complètes*, vol.21, p.446 (my translation).
6 Christiaan Huygens, *Treatise on Light*, trans. Sylvanus P. Thompson (London: Macmillan, 1912), p.3.
7 Both in *Rohault's System of Natural Philosophy, Illustrated with Dr. Samuel Clarke's Notes* (London, 1723), an English version of the *Traité de physique* of 1671, vol.1, author's preface, p.A6r. Cf. above, Chapter 3, section III, p.60 for Bacon's remarks.
8 *Rohault's System*, vol.1, p.b1r.
9 Ibid., p.115.
10 A celebrated work by Bernard Bouvier de Fontenelle, *Entretiens sur la pluralité des mondes* ("Conversations on the Plurality of Worlds," 1686), presents fictional, polite conversations about natural philosophy between a philosopher (the author) and a young lady. The cosmology is Cartesian, and the cultural style is very much that of the salons.
11 Isaac Newton, *The Principia: Mathematical Principles of Natural Philosophy*, trans. I. Bernard Cohen and Anne Whitman (Berkeley, etc.: University of California Press, 1999), p.444. The term "centripetal force" ("centre-seeking" force) was Newton's coinage, intended as a correlate to Huygens's already-published term "centrifugal force" ("centre-fleeing" force).
12 Newton, *Principia*, p.588.
13 Isaac Newton, *Opticks, or A Treatise of the Reflections, Refractions, Inflections & Colours of Light* (New York: Dover, 1952), p.376 (Qu.31), as found in the 1717 third edition, translating a passage in the 1706 second (Latin) edition (the first edition, in English, appeared in 1704).
14 Quoted in Alexandre Koyré, *From the Closed World to the Infinite Universe* (Baltimore: The Johns Hopkins University Press, 1957), p.178.
15 Newton, *Principia*, pp.941–2.
16 Newton, *Opticks*, p.403.
17 Quoted in Margaret C. Jacob, *The Newtonians and the English Revolution, 1689–1720* (Ithaca: Cornell University Press, 1976), p.156.
18 Quoted in Alexandre Koyré, *Newtonian Studies* (Chicago: University of Chicago Press, 1965), p.115.

19 Newton, *Principia*, p.393.
20 Recall too that it was Clarke's Latin edition (1697), with its copious Newtonian annotations, of Rohault's Cartesian textbook of 1671 that was translated into English in 1723 as *Rohault's System of Natural Philosophy*.

Conclusion

1 See Chapter 3, section II, above.

Documentation and
Further Reading

The literature on the so-called Scientific Revolution is enormous. The following material serves as fuller documentation to individual chapters, indicating some of the chief secondary sources to which my own text is indebted in addition to those cited in the notes. The references given will also serve readers who wish to investigate in greater depth any of the issues with which my account deals. With very few exceptions, all the literature cited is in English.

Three recent accounts of the period giving a variety of different perspectives are James R. Jacob, *The Scientific Revolution: Aspirations and Achievements, 1500–1700* (Atlantic Highlands, N.J.: Humanities Press, 1998); John Henry, *The Scientific Revolution and the Origins of Modern Science* (London: Macmillan, 1997), with valuable annotations to its many bibliographical references; and Steven Shapin, *The Scientific Revolution* (Chicago: University of Chicago Press, 1996), containing bibliographical essays on major historiographical themes. Lisa Jardine, *Ingenious Pursuits: Building the Scientific Revolution* (London: Little, Brown, 1999) is a readable general account of the period, while at a more comprehensive level, H. Floris Cohen, *The Scientific Revolution: A Historiographical Inquiry* (Chicago: University of Chicago Press, 1994), deals at length with the historiography prior to about 1990. Among many older studies, Edwin Arthur Burtt, *The Metaphysical Foundations of Modern Physical Science* (Garden City, N.Y.: Doubleday Anchor, 1954 [1932]) is of particular influence on the present work.

Introduction

On the eighteenth-century concept of the Scientific Revolution, and much else, see I. Bernard Cohen, *Revolution in Science* (Cambridge, Mass.: Harvard University Press, 1985). For an excellent overview of recent trends in the history of science, see Jan Golinski, *Making Natural Knowledge: History of Science after Constructivism* (Cambridge: Cambridge University Press, 1998). On historiographical issues in general, a useful recent text is Beverley C. Southgate, *History, What and Why? Ancient, Modern, and Postmodern Perspectives* (London: Routledge, 1996). For those interested in the period immediately preceding that discussed in the present book, Edward Grant, *The Foundations of Modern Science in the Middle Ages: Their Religious, Institutional, and Intellectual Contexts* (Cambridge: Cambridge University Press, 1996), is an up-to-date overview that looks towards the Scientific Revolution. Aristotle's natural philosophy is best approached via the classic by G. E. R. Lloyd, *Aristotle: The Growth and Structure of His Thought* (Cambridge: Cambridge University Press, 1968). Literature on Francis Bacon is considered below, under Chapter 3. On the scope of discovery metaphors in this period, a suggestive article is Amir Alexander, "The Imperialist Space of Elizabethan Mathematics," *Studies in History and Philosophy of Science* 26 (1995), pp.559–91.

Chapter 1

Grant, *The Foundations of Modern Science in the Middle Ages*, is the best general introduction on the Middle Ages; it may be usefully supplemented by the same author's suggestive article, Edward Grant, "Aristotelianism and the Longevity of the Medieval World View," *History of Science* 16 (1978), pp.93–106. David F. Noble, *A World Without Women: The Christian Clerical Culture of Western Science* (New York: Knopf, 1992), explains some of the social realities of the medieval and early-modern universities in an accessible style.

Barry Barnes, "On the Conventional Character of Knowledge and Cognition," in Karin Knorr-Cetina and Michael Mulkay (ed.), *Science Observed: Perspectives on the Social Study of Science* (London: Sage, 1983), pp.19–51, explains some of the basic conceptions on which "constructivist" approaches to science, which emphasize the socio-cultural shaping of scientific knowledge, are based. On the issue of whether the "Scientific Revolution" is properly named, see Stephen Pumfrey, "No Science, Therefore No Scientific Revolution? Social Constructionist Approaches to 16th and 17th Century Studies of Nature," in Dominique Pestre (ed.), *L'étude sociale des sciences* (Paris: Centre de Recherche en Histoire des Sciences et des Techniques, 1992), pp.61–86. On the category of "natural philosophy" and its difference from the modern category of "science," see Andrew Cunningham and Roger French, *Before Science: The Invention of the Friar's Natural Philosophy* (Aldershot: Scolar Press, 1996), and Andrew Cunningham, "How the Principia Got its Name; or, Taking Natural Philosophy Seriously," *History of Science* 29 (1991), pp.377–92.

Giovanna Ferrari, "Public Anatomy Lessons and the Carnival: The Anatomy Theatre of Bologna," *Past and Present*, no. 117 (1987), pp.50–106, C. D. O'Malley, *Andreas Vesalius of Brussels 1514–1564* (Berkeley, etc.: University of California Press, 1964), and Jerome J. Bylebyl, "Interpreting the 'Fasciculo' Anatomy Scene," *Journal of the History of Medicine and Allied Sciences* 45 (1990), pp.285–316, provide various perspectives on the world of sixteenth-century anatomical practice. On the background to astronomy in this period, Liba C. Taub, *Ptolemy's Universe: The Natural, Philosophical and Ethical Foundations of Ptolemy's Astronomy* (Chicago: Open Court, 1993), Owen Gingerich, "The Accuracy of the Toledan Tables," in *PRISMATA: Festschrift für Willy Hartner*, ed. Y. Maeyama and W. G. Saltzer (Wiesbaden: Steiner, 1977), pp.151–63, and the classic Thomas S. Kuhn, *The Copernican Revolution: Planetary Astronomy in the Development of Western Thought* (Cambridge, Mass.: Harvard University Press, 1957). Edward Grant, "Celestial Orbs in the Latin Middle Ages," *Isis* 78 (1987), pp.153–73, provides important discussion of the legacy of the Middle Ages to sixteenth-century views on the relationship of the natural philosophy of the heavens to mathematical astronomy, which contextualizes the discussions of Nicholas Jardine, "The Significance of the Copernican Orbs", *Journal for the History of Astronomy* 13 (1982), pp.168–94, and Robert S. Westman, "The Astronomer's Role in the Sixteenth Century: A Preliminary Study," *History of Science* 18 (1980), pp.105–47. The technical details of classical astronomy are impressively introduced by James Evans, *The History and Practice of Ancient Astronomy* (New York: Oxford University Press, 1998).

On the character of Aristotelian natural philosophy and its critics, see Keith Hutchison, "Dormitive Virtues, Scholastic Qualities, and the New Philosophies," *History of Science* 29 (1991), pp.245–78. The role of printing in restructuring perceptions of knowledge during the Scientific Revolution are discussed in Adrian Johns, *The Nature of the Book: Print and Knowledge in the Making* (Chicago: University of Chicago Press, 1998), and the classic Elizabeth L. Eisenstein, *The Printing Press as an Agent of Change: Communications and Cultural Transformations in Early-Modern Europe*, 2 vols (Cambridge: Cambridge University Press, 1980).

Important discussions that inaugurated the close examination of magical traditions in the origins of modern science include Frances A. Yates, *Giordano Bruno and the Hermetic Tradition* (Chicago: University of Chicago Press, 1979 [1964]); Yates, "The Hermetic Tradition in Renaissance Science," in Charles S. Singleton (ed.), *Art, Science and History in the Renaissance* (Baltimore: The Johns Hopkins University Press, 1968), pp.255–74; Eugenio Garin, "Magic and Astrology in the Civilization of the Renaissance," in Garin, *Science and Civic Life in the Italian Renaissance* (Garden City, N.Y.: Doubleday Anchor Books, 1969), pp.145–65. A more recent overview and critique is Brian Copenhaver, "Natural Magic, Hermetism, and Occultism in Early Modern Science," in *Reappraisals of the Scientific Revolution*, ed. David C. Lindberg and Robert S. Westman (Cambridge: Cambridge University Press, 1990), pp.261–301; see also William Eamon, *Science and the Secrets of Nature: Books of Secrets in Medieval and Early Modern Culture* (Princeton: Princeton University Press, 1994).

For studies of alchemy in the seventeenth century, see Betty Jo Teeter Dobbs, *The Foundations of Newton's Alchemy, or "The Hunting of the Greene Lyon"* (Cambridge: Cambridge University Press, 1975); Dobbs, *The Janus Face of Genius: The Role of Alchemy in Newton's Thought* (Cambridge: Cambridge University Press, 1991); and more recently William R. Newman, *Gehennical Fire: The Lives of George Starkey, an American Alchemist in the Scientific Revolution* (Cambridge, Mass.: Harvard University Press, 1994), and Lawrence Principe, *The Aspiring Adept: Robert Boyle and His Alchemical Quest, including Boyle's "Lost" Dialogue on the Transmutation of Metals* (Princeton: Princeton University Press, 1998).

Cabalism is treated in Gershom Scholem, *Kabbalah* (New York: Meridian, 1978); its specific role in the life of the Elizabethan magus John Dee is considered in Deborah Harkness, *John Dee's Conversations With Angels: Cabala, Alchemy, and the End of Nature* (Cambridge: Cambridge University Press, 1999). On early-modern astrology, see Patrick Curry, *Prophecy and Power: Astrology in Early Modern England* (Princeton: Princeton University Press, 1989); also Patrick Curry (ed.), *Astrology, Science, and Society: Historical Essays* (Woodbridge, England: Boydell Press, 1987).

Chapter 2

On the academic background to the period, David L. Wagner (ed.), *The Seven Liberal Arts in the Middle Ages* (Bloomington: Indiana University Press, 1983), discusses the foundations of medieval learning. Two articles by Paul Oskar Kristeller, "The Humanist Movement" and "Humanism and Scholasticism in the Italian Renaissance," both in Kristeller, *Renaissance Thought: The Classic, Scholastic, and Humanist Strains* (New York: Harper Torchbooks, 1961), pp.3–23 and pp.92–119, are classic introductions to the modern historical understanding of renaissance humanism, as also is Jerrold E. Seigel, *Rhetoric and Philosophy in Renaissance Humanism: The Union of Eloquence and Wisdom* (Princeton: Princeton University Press, 1968).

Jill Kraye (ed.), *The Cambridge Companion to Renaissance Humanism* (Cambridge: Cambridge University Press, 1996), and Anthony Grafton and Lisa Jardine, *From Humanism to the Humanities: Education and the Liberal Arts in Fifteenth- and Sixteenth-Century Europe* (Cambridge, Mass.: Harvard University Press, 1986), are good places to start in understanding the general impact of humanist pedagogy. On science, humanism, and the renaissance, see Brian Copenhaver, "Did Science have a Renaissance?", *Isis* 83 (1992), pp.387–407; Vivian Nutton, "Greek Science in the Sixteenth-Century Renaissance," in J. V. Field and Frank A. J. L. James (eds), *Renaissance and Revolution: Humanists, Scholars, Craftsmen and Natural Philosophers in Early Modern Europe* (Cambridge: Cambridge University Press, 1993), pp.15–28. More disciplinarily-specific studies include Paul Lawrence Rose, *The Italian Renaissance of Mathematics: Studies on Humanists and Mathematicians from Petrarch to Galileo* (Geneva: Droz, 1975); Robert S. Westman, "Proof, Poetics, and Patronage: Copernicus's Preface to *De revolutionibus*," in David C. Lindberg and

Robert S. Westman (eds), *Reappraisals of the Scientific Revolution* (Cambridge: Cambridge University Press, 1990), pp.167–205; Karen Reeds, "Renaissance Humanism and Botany," *Annals of Science* 33 (1976), pp.519–42. Peter Dear, *Discipline and Experience: The Mathematical Way in the Scientific Revolution* (Chicago: University of Chicago Press, 1995), pp.115–23, contains further discussion and references.

William Eamon, *Science and the Secrets of Nature: Books of Secrets in Medieval and Early Modern Culture* (Princeton: Princeton University Press, 1994), is an important study of a significant genre of the period relating to natural knowledge.

Several recent studies stress the humanistic aspects of renaissance anatomy: Andrew Cunningham, *The Anatomical Renaissance: The Resurrection of the Anatomical Projects of the Ancients* (Aldershot: Scolar Press, 1997); R. K. French, *Dissection and Vivisection in the European Renaissance* (Aldershot: Ashgate, 1999); Andrew Wear, R. K. French and I. M. Lonie (eds), *The Medical Renaissance of the Sixteenth Century* (Cambridge: Cambridge University Press, 1985). For more on mathematics and method, JoAnn S. Morse, "The Reception of Diophantus' 'Arithmetic' in the Renaissance" (Ph.D., Princeton University, 1981); Jaako Hintikka and Unto Remes, *The Method of Analysis: Its Geometrical Origin and its General Significance* (Boston Studies in the Philosophy of Science, vol.25) (Dordrecht: Reidel, 1974).

Kuhn, *The Copernican Revolution*, is basic for an introduction to sixteenth-century astronomy. In addition to Jardine, "Significance of the Copernican Orbs," see also Robert S. Westman, "The Melanchthon Circle, Rheticus, and the Wittenberg Interpretation of the Copernican Theory," *Isis* 66 (1975), pp.165–93; Westman, "The Copernicans and the Churches," in *God and Nature: Historical Essays on the Encounter between Christianity and Science*, ed. David C. Lindberg and Ronald L. Numbers (Berkeley, etc.: University of California Press, 1986), pp.76–113. Sachiko Kusukawa, *The Transformation of Natural Philosophy: The Case of Philip Melanchthon* (Cambridge: Cambridge University Press, 1995) and Charlotte Methuen, *Kepler's Tübingen: Stimulus to a Theological Mathematics* (Aldershot: Ashgate, 1998) are important discussions of a specifically Lutheran context for natural philosophical and astronomical work in the sixteenth century; for a more general overview of religious dimensions of early-modern science, see John Hedley Brooke, *Science and Religion: Some Historical Perspectives* (Cambridge: Cambridge University Press, 1991), chs.2–4.

Chapter 3

On Roger Bacon: Jeremiah Hackett, "Roger Bacon on 'scientia experimentalis'," in Jeremiah Hackett (ed.), *Roger Bacon and the Sciences: Commemorative Essays* (Leiden: Brill, 1997). A good general account of the medicine of the Middle Ages may be found in David C. Lindberg, *The Beginnings of Western Science: The European Scientific Tradition in Philosophical, Religious, and Institutional Context, 600 BC. to AD. 1450* (Chicago: University of Chicago Press, 1992), and especially in Nancy G. Siraisi, *Medieval and Early Renaissance Medicine: An Introduction to Knowledge and Practice* (Chicago: University of Chicago Press, 1990).

On Paracelsus and Paracelsianism: Charles Webster, *From Paracelsus to Newton: Magic and the Making of Modern Science* (Cambridge: Cambridge University Press, 1982); Walter Pagel, *Paracelsus: An Introduction to Philosophical Medicine in the Era of the Renaissance*, 2nd rev. edn (Basel and New York: Karger, 1982); Andrew Weeks, *Paracelsus: Speculative Theory and the Crisis of the Early Reformation* (Albany: State University of New York Press, 1997); Allen G. Debus, *The French Paracelsians: The Chemical Challenge to Medical and Scientific Tradition in Early Modern France* (Cambridge: Cambridge University Press, 1991); Allen G. Debus, *The English Paracelsians* (London: Oldbourne, 1965). Other aspects of the practical dimensions of the study of nature are Pamela H. Smith, *The Business of Alchemy: Science and Culture in the Holy Roman Empire* (Princeton: Princeton University Press, 1994); Owen Hannaway, "Georgius Agricola as Humanist," *Journal of the History of Ideas* 53 (1992), pp.553–60; and the excellent Paolo Rossi,

Philosophy, Technology, and the Arts in the Early Modern Era, trans. Salvator Attanasio (New York: Harper & Row, 1970).

The cultural forms of natural knowledge are considered in Owen Hannaway, "Laboratory Design and the Aim of Science: Andreas Libavius versus Tycho Brahe," *Isis* 77 (1986), pp.585–610; Steven Shapin, " 'The Mind is Its Own Place': Science and Solitude in Seventeenth-Century England," *Science in Context* 4 (1991), pp.191–218; Eamon, *Science and the Secrets of Nature*. On Libavius and Paracelsianism, Owen Hannaway, *The Chemists and the Word: The Didactic Origins of Chemistry* (Baltimore: Johns Hopkins University Press, 1975). Katherine Park and Lorraine Daston, *Wonders and the Order of Nature, 1150–1750* (New York: Zone Books, 1998), is an impressive study of the sensibilities brought to natural phenomena over the course of an extended period in Europe.

On Francis Bacon and his immediate context, Gad Freudenthal, "Theory of Matter and Cosmology in William Gilbert's *De Magnete*," *Isis* 74 (1983), pp.22–37; Edgar Zilsel, "The Origins of William Gilbert's Scientific Method," *Journal of the History of Ideas* 2 (1941), pp.1–32; Julian Martin, *Francis Bacon, the State, and the Reform of Natural Philosophy* (Cambridge: Cambridge University Press, 1992); J. R. Ravetz, "Francis Bacon and the Reform of Natural Philosophy," in *Science, Medicine, and Society in the Renaissance*, 2 vols, ed. Allen G. Debus (New York: Science History Publications, 1972), vol.2, pp.97–119; Graham Rees, "Francis Bacon's Semi-Paracelsian Cosmology," *Ambix* 22 (1975), pp.81–101, 161–73. The monumental work on the subsequent career of Bacon's proposals is Charles Webster, *The Great Instauration: Science, Medicine, and Reform 1626–1660* (London: Duckworth, 1975).

Chapter 4

Methodological and intellectual contexts for Galileo's work are discussed in Nicholas Jardine, "Epistemology of the Sciences," in *The Cambridge History of Renaissance Philosophy*, ed. Charles Schmitt, Quentin Skinner, Eckhard Kessler and Jill Kraye (Cambridge: Cambridge University Press, 1988), pp.685–711. Dear, *Discipline and Experience*, considers Jesuit colleges and the teaching of mathematical sciences, as does William A. Wallace, *Galileo and His Sources: The Heritage of the Collegio Romano in Galileo's Science* (Princeton: Princeton University Press, 1984). For more on the Jesuit tradition, Steven J. Harris, "Transposing the Merton Thesis: Apostolic Spirituality and the Establishment of the Jesuit Scientific Tradition," *Science in Context* 3 (1989), pp.29–65.

Of very many works on Galileo, the following are of particular use and relevance here: Stillman Drake, *Galileo at Work: His Scientific Biography* (Chicago: University of Chicago Press, 1978); Michael Sharratt, *Galileo: Decisive Innovator* (Cambridge: Cambridge University Press, 1994); Maurice Clavelin, *The Natural Philosophy of Galileo: Essay on the Origins and Formation of Classical Mechanics*, trans. A. J. Pomerans (Cambridge, Mass.: MIT Press, 1974); Stillman Drake and I. E. Drabkin (eds), *Mechanics in Sixteenth-Century Italy: Selections from Tartaglia, Benedetti, Guido Ubaldo, and Galileo* (Madison: University of Wisconsin Press, 1969) for early Galilean writings; Martha Fehér, "Galileo and the Demonstrative Ideal of Science," *Studies in History and Philosophy of Science* 13 (1982), pp.87–110 considers the relation of Galileo's mathematical work to issues of essences in natural philosophy.

Stillman Drake (ed. and trans.), *Discoveries and Opinions of Galileo* (Garden City, N.Y.: Doubleday Anchor, 1957), contains much translated material with commentary; William R. Shea, *Galileo's Intellectual Revolution* (London: Macmillan, 1972), is especially useful on Galileo's earlier controversies. Mario Biagioli, *Galileo, Courtier: The Practice of Science in the Culture of Absolutism* (Chicago: University of Chicago Press, 1993), is a valuable perspective on Galileo's work in the context of his patronage strategies. More generally, Shapin, *The Scientific Revolution*, chapter 3, considers the goals and settings for scientific activities in this period.

On Tycho and Kepler, see Victor E. Thoren, *The Lord of Uraniborg: A Biography of Tycho Brahe* (Cambridge: Cambridge University Press, 1990); Max Caspar, *Kepler*, trans. C. Doris Hellman, new edition with notes by Owen Gingerich (New York: Dover, 1993); Nicholas Jardine, *The Birth of History and Philosophy of Science: Kepler's "A Defence of Tycho against Ursus" with Essays on Its Provenance and Significance* (Cambridge: Cambridge University Press, 1984); J. V. Field, *Kepler's Geometrical Cosmology* (London: Athlone, 1988); Bruce Stephenson, *Kepler's Physical Astronomy* (Princeton: Princeton University Press, 1987); Robert S. Westman, "Three Responses to the Copernican Theory: Johannes Praetorius, Tycho Brahe, and Michael Maestlin," in Robert S. Westman (ed.), *The Copernican Achievement* (Berkeley, etc.: University of California Press, 1975), pp.285–345.

Siraisi, *Medieval and Early Renaissance Medicine*, touches on medical astrology; for a slightly different perspective on this, see also Lynn White, "Medical Astrologers and Late Medieval Astrology," *Viator* 6 (1975), pp.295–308.

Mathematical practitioners in England are discussed in the classic E. G. R. Taylor, *The Mathematical Practitioners of Tudor and Stuart England* (Cambridge: Cambridge University Press, 1954). More recently, see J. A. Bennett, "The Mechanics' Philosophy and the Mechanical Philosophy," *History of Science* 24 (1986) pp.1–28; J. A. Bennett, "The Challenge of Practical Mathematics," in *Science, Culture and Popular Belief in Renaissance Europe*, ed. Stephen Pumfrey, Paolo L. Rossi and Maurice Slawinski (Manchester: Manchester University Press, 1991), pp.176–90; Lesley B. Cormack, *Charting an Empire: Geography at the English Universities* (Chicago: University of Chicago Press, 1997); Stephen Johnston, "Mathematical Practitioners and Instruments in Elizabethan England," *Annals of Science* 48 (1991), pp.319–44; Stephen Johnston, "The Identity of the Mathematical Practitioner in 16th-Century England," in Irmgard Hantsche (ed.), *Der "mathematicus": Zur Entwicklung und Bedeutung einer neuen Berufsgruppe in der Zeit Gerhard Mercators* (Bochum: Brockmeyer, 1996), pp.93–120; Katherine Hill, " 'Juglers or Schollers?': Negotiating the Role of a Mathematical Practitioner," *British Journal for the History of Science* 31 (1998), pp.253–74.

Chapter 5

Recent biographies of Descartes are Stephen Gaukroger, *Descartes: An Intellectual Biography* (Oxford: Clarendon Press, 1995); William R. Shea, *The Magic of Numbers and Motion: The Scientific Career of René Descartes* (New York: Science History Publications, 1991), with especial concentration on the scientific work; Daniel Garber, *Descartes' Metaphysical Physics* (Chicago: University of Chicago Press, 1992); Geneviève Rodis-Lewis, *Descartes: His Life and Thought*, trans. Jane Marie Todd (Ithaca: Cornell University Press, 1998).

Descartes's contemporaries are discussed in R. Hooykaas, "Beeckman, Isaac," in Charles C. Gillispie (ed.), *Dictionary of Scientific Biography*, vol.1 (New York: Scribner's, 1970), pp.566–8; Richard H. Popkin, *The History of Scepticism from Erasmus to Spinoza* (Berkeley, etc.: University of California Press, 1979); Lynn Sumida Joy, *Gassendi the Atomist: Advocate of History in an Age of Science* (Cambridge: Cambridge University Press, 1987).

Specific aspects of Descartes's work are analysed in Bruce S. Eastwood, "Descartes on Refraction: Scientific versus Rhetorical Method," *Isis* 75 (1984), pp.481–502; A. Mark Smith, "Descartes's Theory of Light and Refraction: A Discourse on Method," *Transactions of the American Philosophical Society* 77, Part 3 (1987); Peter Galison, "Descartes' Comparisons: From the Visible to the Invisible," *Isis* 75 (1984), pp.311–26; Étienne Gilson, "Météores cartésiens et météores scolastiques," in Gilson, *Études sur le rôle de la pensée médiévale dans la formation du système cartésien* (Paris: J. Vrin, 1930), pp.102–37; Desmond Clarke, *Descartes' Philosophy of Science* (Manchester: Manchester University Press, 1982). Intellectual context for Descartes's hugely expanded universe is provided by Steven J. Dick, *Plurality of Worlds: The Origins of the*

Extraterrestrial Life Debate from Democritus to Kant (Cambridge: Cambridge University Press, 1982).

A classic account of English corpuscular mechanism is Robert Kargon, *Atomism in England from Hariot to Newton* (Oxford: Clarendon Press, 1966). The mechanical philosophy of Robert Boyle has been recently considered in the context of his experimentalism and hypotheticalism by Rose-Mary Sargent, *The Diffident Naturalist: Robert Boyle and the Philosophy of Experiment* (Chicago: University of Chicago Press, 1995).

Chapter 6

On the social and institutional places available in the early-modern period for philosophers of nature, see Westman, "Astronomer's Role"; essays in Bruce T. Moran (ed.), *Patronage and Institutions: Science, Technology, and Medicine at the European Court 1500–1750* (Woodbridge, Suffolk: The Boydell Press, 1991); John Gascoigne, "A Reappraisal of the Role of the Universities in the Scientific Revolution," in David C. Lindberg and Robert S. Westman (eds), *Reappraisals of the Scientific Revolution* (Cambridge: Cambridge University Press, 1990), pp.207–60. Biagioli, *Galileo, Courtier*; Richard S. Westfall, "Science and Patronage: Galileo and the Telescope," *Isis* 76 (1985), pp.11–30; Mario Biagioli, "The Social Status of Italian Mathematicians, 1450–1600," *History of Science* 27 (1989), pp.41–95, all consider this issue with especial reference to Galileo.

On Galileo's conflicts with the Catholic church, see for a general discussion Peter Dear, "The Church and the New Philosophy," in Stephen Pumfrey, Paolo Rossi and Maurice Slawinski (eds), *Science, Culture and Popular Belief in Early Modern Europe* (Manchester: Manchester University Press, 1991), pp.119–39. There are several accounts of the details of Galileo's difficulties, including the classic account by Giorgio De Santillana, *The Crime of Galileo* (Chicago: University of Chicago Press, 1955); also Richard J. Blackwell, *Galileo, Bellarmine, and the Bible* (Notre Dame: University of Notre Dame Press, 1991); Jerome J. Langford, *Galileo, Science, and the Church*, 3rd edn (Ann Arbor: University of Michigan Press, 1992); Rivka Feldhay, *Galileo and the Church: Political Inquisition or Critical Dialogue?* (Cambridge: Cambridge University Press, 1995). Maurice A. Finocchiaro (ed.), *The Galileo Affair: A Documentary History* (Berkeley, etc.: University of California Press, 1989) is a useful source book.

On John Dee, see Nicholas H. Clulee, *John Dee's Natural Philosophy: Between Science and Religion* (London: Routledge, 1988); Harkness, *John Dee's Conversations With Angels*. For other patronage beneficiaries in England, John Shirley, *Thomas Harriot: A Biography* (Oxford: Clarendon Press, 1983); Tom Sorell, *Hobbes* (London: Routledge & Kegan Paul, 1986).

Biagioli, "Scientific Revolution, Social Bricolage, and Etiquette," in Porter and Teich, *The Scientific Revolution in National Context*, touches on a number of different national settings.

On Harvey, see especially Roger French, *William Harvey's Natural Philosophy* (Cambridge: Cambridge University Press, 1994); other useful material appears in Kenneth J. Franklin, "Introduction," in Franklin (ed.), *William Harvey: The Circulation of the Blood and Other Writings* (London: Everyman, 1963); Gweneth Whitteridge, *William Harvey and the Circulation of the Blood* (London: Macdonald, 1971); Walter Pagel, *William Harvey's Biological Ideas: Selected Aspects and Historical Background* (New York: Hafner, 1967); Walter Pagel, *New Light on William Harvey* (Basel: Karger, 1976). An important study of English philosophical comportment in this period is Steven Shapin, *A Social History of Truth: Civility and Science in Seventeenth-Century England* (Chicago: University of Chicago Press, 1994).

Useful insights on the Accademia dei Lincei appear in Pietro Redondi, *Galileo Heretic*, trans. Raymond Rosenthal (Princeton: Princeton University Press, 1987), and Paula Findlen, *Possessing Nature: Museums, Collecting, and Scientific Culture in Early Modern Italy* (Berkeley, etc.:

University of California Press, 1994). On a later Italian academy, see W. E. Knowles Middleton, *The Experimenters: A Study of the Accademia del Cimento* (Baltimore: The Johns Hopkins University Press, 1971), Biagioli, "Scientific Revolution," and Jay Tribby, "Cooking (with) Clio and Cleo: Eloquence and Experiment in Seventeenth-Century Florence," *Journal of the History of Ideas* 52 (1991), pp.417–39.

The Parisian Royal Academy of Sciences has been examined at length in Roger Hahn, *The Anatomy of a Scientific Institution: The Paris Academy of Sciences, 1666–1803* (Berkeley, etc.: University of California Press, 1971); Alice Stroup, *A Company of Scientists: Botany, Patronage, and Community at the Seventeenth-Century Parisian Royal Academy of Sciences* (Berkeley, etc.: University of California Press, 1990). More generally on French scientific academies in the seventeenth century, David S. Lux, "The Reorganization of Science, 1450–1700," in Bruce T. Moran (ed.), *Patronage and Institutions: Science, Technology, and Medicine at the European Court, 1500–1750* (Rochester, N.Y.: Boydell, 1991), pp.185–94; David S. Lux, "Societies, Circles, Academies, and Organizations: A Historiographic Essay on Seventeenth-Century Science," in Peter Barker and Roger Ariew (eds), *Revolution and Continuity: Essays in the History and Philosophy of Early Modern Science* (Washington, D.C.: Catholic University of America Press, 1991), pp.23–43; David S. Lux, *Patronage and Royal Science in Seventeenth-Century France: The Académie de Physique in Caen* (Ithaca: Cornell University Press, 1989). The character of much of the work done in the early Academy is discussed by Christian Licoppe, "The Crystallization of a New Narrative Form in Experimental Reports (1660–1690): Experimental Evidence as a Transaction Between Philosophical Knowledge and Aristocratic Power," *Science in Context* 7 (1994), pp.205–44.

On the Royal Society of London: Michael Hunter, *Science and Society in Restoration England* (Cambridge: Cambridge University Press, 1981); Margery Purver, *The Royal Society: Concept and Creation* (London: Routledge & Kegan Paul, 1967), but note the serious criticisms of Charles Webster, review of Margery Purver, *The Royal Society*, in *History of Science* 6 (1967), pp.106–28.

Elsewhere in the British Isles, K. Theodore Hoppen, *The Common Scientist in the Seventeenth Century: A Study of the Dublin Philosophical Society, 1683–1708* (London: Routledge & Kegan Paul, 1970); much material in Webster, *Great Instauration*; also in Shapin, *Social History of Truth*. The Oxford group of physiologists in the 1650s is examined in Robert G. Frank, Jr., *Harvey and the Oxford Physiologists: Scientific Ideas and Social Interaction* (Berkeley, etc.: University of California Press, 1980). For important light on Boyle, see William R. Newman, "The Alchemical Sources of Robert Boyle's Corpuscular Philosophy," *Annals of Science* 53 (1996), pp.567–85. On Hooke's most famous work, see Michael Aaron Dennis, "Graphic Understanding: Instruments and Interpretation in Robert Hooke's *Micrographia*," *Science in Context* 3 (1989), pp.309–64, and John T. Harwood, "Rhetoric and Graphics in *Micrographia*," in Michael Hunter and Simon Schaffer (eds), *Robert Hooke: New Studies* (Woodbridge, Suffolk: The Boydell Press, 1989), pp.119–47. Carolyn Merchant, *The Death of Nature: Women, Ecology and the Scientific Revolution* (New York: HarperCollins, 1990), chapter 11, discusses Anne Conway, her philosophy and its influence, Margaret Cavendish, and other issues regarding the role of women as participants in and audiences for natural philosophy in the later seventeenth century and early eighteenth century; for more on Cavendish and the Royal Society, see Anna Battigelli, *Margaret Cavendish and the Exiles of the Mind* (Lexington, Ky.: University Press of Kentucky, 1998), chapter 5. See also in general Londa Schiebinger, *The Mind Has No Sex? Women in the Origins of Modern Science* (Cambridge, Mass.: Harvard University Press, 1989).

On the home as a locus for philosophical activity, see Deborah E. Harkness, "Managing an Experimental Household: The Dees of Mortlake and the Practice of Natural Philosophy," *Isis* 88 (1997), pp.247–62; Steven Shapin, "The House of Experiment in Seventeenth-Century England," *Isis* 79 (1988), pp.373–404. On "invisible technicians," Shapin, *Social History of Truth*, chapter 8.

The institutional power of the Jesuits is the subject of Steven J. Harris, "Confession-

Building, Long-Distance Networks, and the Organization of Jesuit Science," *Early Science and Medicine* 1 (1996), pp.287–318; see also Jonathan Spence, *The Memory Palace of Matteo Ricci* (New York: Viking Penguin, 1984).

On natural history collections, see Findlen, *Possessing Nature*. On European attitudes to the peoples of the New World, the classic treatment is Lewis Hanke, *Aristotle and the American Indians: A Study in Race Prejudice in the Modern World* (London: Hollis & Carter, 1959); see also Anthony Pagden, *European Encounters with the New World: From Renaissance to Romanticism* (New Haven: Yale University Press, 1993). Two useful approaches to natural history and its significance are Harold J. Cook, "The New Philosophy and Medicine in Seventeenth-Century England," emphasizing the medical side of the Scientific Revolution, and William B. Ashworth, Jr., "Natural History and the Emblematic World View," both in David C. Lindberg and Robert S. Westman (eds), *Reappraisals of the Scientific Revolution* (Cambridge: Cambridge University Press, 1990), pp.397–436, 303–32, respectively, while the best introduction to the entire subject is provided by the essays by Ashworth, Cunningham, Findlen, Whitaker, Cook, and Johns, in Part I of N. Jardine, J. A. Secord and E. C. Spary (eds), *Cultures of Natural History* (Cambridge: Cambridge University Press, 1996); see also the survey in Allen G. Debus, *Man and Nature in the Renaissance* (Cambridge: Cambridge University Press, 1978), chapter 3. On the establishment of the Jardin du Roi (Jardin des Plantes) in Paris, Rio C. Howard, "Guy de La Brosse and the Jardin des Plantes in Paris," in Harry Woolf (ed.), *The Analytic Spirit: Essays in the History of Science in Honor of Henry Guerlac* (Ithaca: Cornell University Press, 1981), pp.195–224, and Rio C. Howard, *La bibliothèque et le laboratoire de Guy de La Brosse au Jardin des Plantes à Paris* (Geneva: Droz, 1983), are basic sources.

Later seventeenth-century philosophical conceptions of natural history and classification are discussed in Mary M. Slaughter, *Universal Languages and Scientific Taxonomy in the Seventeenth Century* (Cambridge: Cambridge University Press, 1982); see also Phillip R. Sloan, "John Locke, John Ray, and the Problem of the Natural System," *Journal of the History of Biology* 5 (1972), pp.1–53.

Chapter 7

A classic starting point in understanding philosophical conceptualizations of experience is Charles B. Schmitt, "Experience and Experiment: A Comparison of Zabarella's View with Galileo's in *De motu*," *Studies in the Renaissance* 16 (1969), pp.80–138; much else is referenced in Dear, *Discipline and Experience*. In addition, Daniel Garber, "Descartes and Experiment in the *Discourse* and the *Essays*," in Stephen Voss (ed.), *Essays on the Philosophy and Science of René Descartes* (Oxford: Clarendon Press, 1993), and Garber, *Descartes' Metaphysical Physics*, are valuable studies of Descartes's approach to these matters, as is Clarke, *Descartes' Philosophy of Science*. A detailed investigation covering a longer time-span than that considered here is Christian Licoppe, *La formation de la pratique scientifique: Le discours de l'expérience en France et en Angleterre, 1630–1820* (Paris: Éditions La Découverte, 1996).

Alexandre Koyré, "A Documentary History of the Problem of Fall from Kepler to Newton: De motu gravium naturaliter cadentium in hypothesi terrae motae," *Transactions of the American Philosophical Society* n.s.45 (1955), Pt.4, discusses the work of Riccioli. On Torricelli and barometers, see W. E. Knowles Middleton, *The History of the Barometer* (Baltimore: Johns Hopkins University Press, 1964).

In addition to Shapin, *Social History of Truth*, Steven Shapin, " 'A Scholar and a Gentleman': The Problematic Identity of the Scientific Practitioner in Early Modern England," *History of Science* 29 (1991), pp.279–327, also considers the place of philosophy and truth-telling as part of the persona of a philosopher in England. The Royal Society's attitude towards observational and experimental reports as a central part of its enterprise is considered in Peter Dear, "*Totius in verba*: Rhetoric and Authority in the Early Royal Society," *Isis* 76 (1985), pp.145–61;

above all, see the classic Steven Shapin and Simon Schaffer, *Leviathan and the Air-Pump: Hobbes, Boyle, and the Experimental Life* (Princeton: Princeton University Press, 1985).

On Newton: his early studies are presented and discussed in J. E. McGuire and Martin Tamny, *Certain Philosophical Questions: Newton's Trinity Notebook* (Cambridge: Cambridge University Press, 1983). Alan E. Shapiro, *Fits, Passions, and Paroxysms: Physics, Method, and Chemistry and Newton's Theories of Colored Bodies and Fits of Easy Reflection* (Cambridge: Cambridge University Press, 1993), considers in addition Newton's mature optical studies. Simon Schaffer, "Glass Works: Newton's Prisms and the Uses of Experiment," in David Gooding, Trevor Pinch and Simon Schaffer (eds), *The Uses of Experiment: Studies in the Natural Sciences* (Cambridge: Cambridge University Press, 1989), pp.67–104, looks at the reception of Newton's optical ideas, as does Zev Bechler, "Newton's 1672 Optical Controversies: A Study in the Grammar of Scientific Dissent," in Yehuda Elkana (ed.), *The Interaction Between Science and Philosophy* (Atlantic Highlands, N.J.: Humanities Press, 1974), pp.115–42. Alan E. Shapiro, "The Gradual Acceptance of Newton's Theory of Light and Color, 1672–1727," *Perspectives on Science: Historical, Philosophical, Social* 4 (1996), pp.59–140, takes issue with Schaffer on the issue of the primacy of theory.

On Harvey, in addition to the essential French, *Harvey*, see the fascinating article by Andrew Wear, "William Harvey and the 'Way of the Anatomists'," *History of Science* 21 (1983), pp.223–49, which presents Harvey's work as the approach of an anatomist intent on "seeing" rather than on testing hypotheses.

Chapter 8

On Descartes's work in relation to that of his scholastic contemporaries, see Roger Ariew, *Descartes and the Last Scholastics* (Ithaca: Cornell University Press, 1999); the scholastic context for much of the philosophical work of Descartes and many others in this period is valuably surveyed in Christia Mercer, "The Vitality and Importance of Early Modern Aristotelianism," in Tom Sorell (ed.), *The Rise of Modern Philosophy: The Tension Between the New and Traditional Philosophies from Machiavelli to Leibniz* (Oxford: Clarendon Press, 1993), pp.33–67. The early reception of Descartes's ideas in the Netherlands is the subject of Robert S. Westman, "Huygens and the Problem of Cartesianism," in H. J. M. Bos et al. (eds), *Studies on Christiaan Huygens: Invited Papers from the Symposium on the Life and Work of Christiaan Huygens, Amsterdam, 22–25 August 1979* (Lisse: Swets & Zeitlinger, 1980), pp.83–103, and Theo Verbeek, *Descartes and the Dutch: Early Reactions to Cartesian Philosophy, 1637–1650* (Carbondale: Southern Illinois University Press, 1992). The influence of Descartes's ideas in England is the subject of Laurens Laudan, "The Clock Metaphor and Probabilism: The Impact of Descartes on English Methodological Thought, 1650–65," *Annals of Science* 22 (1966), pp.73–104, which is disputed by G. A. J. Rogers, "Descartes and the Method of English Science," *Annals of Science* 29 (1972), pp.237–55. Further investigation of Boyle's methodological views may be found in Jan Wojcik, *Robert Boyle and the Limits of Reason* (Cambridge: Cambridge University Press, 1997).

Huygens's mechanics is studied in Joella G. Yoder, *Unrolling Time: Christiaan Huygens and the Mathematization of Nature* (Cambridge: Cambridge University Press, 1988), while contextualization of his work appears in Geoffrey V. Sutton, *Science for a Polite Society: Gender, Culture, and the Demonstration of Enlightenment* (Boulder, Col.: Westview, 1995). The older English biography, Arthur Bell, *Christiaan Huygens and the Development of Science in the Seventeenth Century* (New York: Longmans Green, 1947), is still of some value as a biographical overview. On Huygens's theory of gravity, see E. J. Aiton, *The Vortex Theory of Planetary Motions* (London: Macdonald, 1972). Huygens's theory of light is discussed in Alan E. Shapiro, "Huygens' Kinematic Theory of Light," in Bos, *Studies on Christiaan Huygens*, pp.200–20.

On Rohault, see Sutton, *Science for a Polite Society*; the classic study on later Cartesianism in France is Paul Mouy, *Le développement de la physique cartésienne, 1646–1712* (Paris: J. Vrin, 1934). L. W. B. Brockliss, "Aristotle, Descartes and the New Science: Natural Philosophy at the University of Paris, 1600–1740," *Annals of Science* 38 (1981), pp.33–69, and Brockliss, *French Higher Education in the Seventeenth and Eighteenth Centuries: A Cultural History* (Oxford: Clarendon Press, 1987), are useful insights into the adoption of Cartesian ideas by French universities.

Erica Harth, *Cartesian Women: Versions and Subversions of Rational Discourse in the Old Regime* (Ithaca: Cornell University Press, 1992), like Sutton, *Science for a Polite Society*, discusses the place of Cartesianism in salon culture and among women, as does Schiebinger, *The Mind Has No Sex?*.

Useful introductions to Leibniz are E. J. Aiton, *Leibniz – A Biography* (Bristol: Adam Hilger, 1989); essays in Nicholas Jolley (ed.), *The Cambridge Companion to Leibniz* (Cambridge: Cambridge University Press, 1995).

On Newton, besides McGuire and Tamny, *Certain Philosophical Questions*, concerning his early work, see the now-standard biography by Richard S. Westfall, *Never at Rest: A Biography of Isaac Newton* (Cambridge: Cambridge University Press, 1980).

The clearest account of the classic view of the Scientific Revolution as proceeding via Kepler and Galileo to Newton is I. Bernard Cohen, *The Birth of a New Physics*, rev. and updated edn (New York: W. W. Norton, 1985), while philosophical and metaphysical dimensions of the story are examined in Alexandre Koyré, *From the Closed World to the Infinite Universe* (Baltimore: Johns Hopkins University Press, 1957). Richard S. Westfall, *Force in Newton's Physics: The Science of Dynamics in the Seventeenth Century* (London: Macdonald, 1971), presents a similar view of Newton in an accessible yet technical style.

The rise of "Newtonianism" in the eighteenth century is the subject of Betty Jo Teeter Dobbs and Margaret C. Jacob, *Newton and the Culture of Newtonianism* (Atlantic Highlands, N. J.: Humanities Press, 1995), chapter 2; Margaret C. Jacob, *The Newtonians and the English Revolution 1689–1720* (Ithaca: Cornell University Press, 1976); also Margaret C. Jacob, "The Truth of Newton's Science and the Truth of Science's History: Heroic Science at its Eighteenth-Century Formulation," in Margaret J. Osler (ed.), *Rethinking the Scientific Revolution* (Cambridge: Cambridge University Press, 2000), pp.315–32; Larry Stewart, *The Rise of Public Science: Rhetoric, Technology, and Natural Philosophy in Newtonian Britain, 1660–1750* (Cambridge: Cambridge University Press, 1992). The dominance of Newton as president of the Royal Society in the early eighteenth century is also examined in Schaffer, "Glass Works"; John L. Heilbron, *Physics at the Royal Society During Newton's Presidency* (Los Angeles: William Andrews Clark Memorial Library, 1983); and in the latter chapters of Marie Boas Hall, *Promoting Experimental Learning: Experiment and the Royal Society 1660–1727* (Cambridge: Cambridge University Press, 1991); see also John L. Heilbron, *Electricity in the Seventeenth and Eighteenth Centuries: A Study in Early Modern Physics* (Berkeley, etc.: University of California Press, 1979).

Aspects of Newton's natural philosophy in this period, and their later repercussions, are discussed in P. M. Heimann [Harman], " 'Nature is a Perpetual Worker': Newton's Aether and Eighteenth-Century Natural Philosophy," *Ambix* 20 (1973), pp.1–25; P. M. Heimann [Harman] and J. E. McGuire, "Newtonian Forces and Lockean Powers: Concepts of Matter in Eighteenth-Century Thought," *Historical Studies in the Physical Sciences* 3 (1971), pp.233–306. Criticisms of Newton's doctrines of space and gravity are examined in Alexandre Koyré, "Huygens and Leibniz on Universal Attraction," in Koyré, *Newtonian Studies* (Chicago: University of Chicago Press, 1965), pp.115–38. On the famous Leibniz–Clarke debate, see Koyré, *Closed World*, chapter 11, for more on the context of which see A. Rupert Hall, *Philosophers at War: The Quarrel Between Newton and Leibniz* (Cambridge: Cambridge University Press, 1980); Domenico Bertoloni Meli, *Equivalence and Priority: Newton versus Leibniz. Including Leibniz's Unpublished Manuscripts on the Principia* (Oxford: Clarendon Press, 1993), and particularly Steven Shapin, "Of Gods and Kings: Natural Philosophy and Politics in the Leibniz–Clarke Disputes," *Isis* 72 (1981), pp.187–215.

Conclusion

Michael S. Mahoney, "Christiaan Huygens: The Measurement of Time and Longitude at Sea," in Bos, *Studies on Christiaan Huygens*, pp.234–70, is a study of one of the most plausible attempts at utilizing theoretical scientific work in the service of practical state interests. On the use of "method" as justification for novel approaches to natural philosophy in the seventeenth century, see Peter Dear, "Method and the Study of Nature," in Daniel Garber and Michael Ayers (eds), *The Cambridge History of Seventeenth-Century Philosophy*, 2 vols, vol.1 (Cambridge: Cambridge University Press, 1998), pp.147–77. Margaret C. Jacob, *Scientific Culture and the Making of the Industrial West* (New York: Oxford University Press, 1997), argues for direct connections between science and the industrial revolution in the eighteenth century.

For the most recent attempt at encompassing the shape of science in the eighteenth century, see now William Clark, Jan Golinski and Simon Schaffer (eds), *The Sciences in Enlightened Europe* (Chicago: University of Chicago Press, 1999).

Dramatis Personae

This list includes most of the individuals mentioned in the text. It is certainly not exhaustive of significant people involved in the sciences during the sixteenth and seventeenth centuries.

An excellent source for detailed biographical and bibliographical entries concerning nearly all of the following, and many others, is Charles C. Gillispie (ed.), *Dictionary of Scientific Biography* (New York: Charles Scribner's Sons, 1970–80).

(All dates are AD except where noted.)

Agricola, Georgius (1494–1555): German author of *De re metallica* (1556), a work on mining and metallurgy.

Aldrovandi, Ulisse (1522–1605): Italian botanist.

Apollonius of Perga (second half of third century to early second century BC): ancient Greek astronomer and mathematician.

Aquinas, Thomas (*c.*1224–1274): Roman Catholic theologian and Aristotelian philosopher.

Archimedes (*c.*287–212 BC): Greek mathematician who wrote on centres of gravity and buoyancy.

Aristotle (384–322 BC): Greek philosopher, of enormous importance for medieval and early-modern universities, who stressed the senses as the source of knowledge.

Averroës [Ibn Rushd] (1126–1198): Arabic commentator on the works of Aristotle.

Avicenna [Ibn Sina] (980–1037): Arabic medical writer and commentator on Aristotle.

Bacon, Francis (1561–1626): English statesman and promoter of natural knowledge as useful for human life.

Bacon, Roger (*c.*1219–*c.*1292): English Franciscan priest who argued for knowledge leading to practical inventions (*scientia experimentalis*).

Baldi, Bernardino (1553–1617): Humanist mathematician and collaborator of Guidobaldo dal Monte.

Beeckman, Isaac (1588–1637): Dutch schoolmaster; a corpuscularian, and an early influence on Descartes.

Bentley, Richard (1662–1742): English follower of Newton.

Bernoulli (later seventeenth century through eighteenth century): a clan of Swiss mathematicians.

Biancani, Giuseppe (1566–1624): Italian Jesuit mathematician.

Biringuccio, Vannuccio (1480–*c.*1539): Italian author of *Pirotechnia* (1540), a work on metallurgy similar to Agricola's (q.v.).

Borelli, Giovanni Alfonso (1608–1679): Italian member of Accademia del Cimento; mathematician and physiologist.

Boyle, Robert (1627–1691): prominent English member of the Royal Society, experimentalist, and promoter of the "mechanical philosophy."

Bruno, Giordano (1548–1600): Italian supporter of unorthodox views about the universe, which included a moving earth and the denial of the Holy Trinity, and burnt by the Catholic Church in Rome for his heresy.

Campanella, Tommaso (1568–1639): Italian author of *City of the Sun* (1623), a utopian work, and a freethinking member of the Catholic Dominican Order who spent much of his life in prison because of his political views.

Cardano, Girolamo (1501–1576): Italian mathematician, philosopher, and astrologer.

Casaubon, Isaac (1559–1614): Huguenot humanist scholar who first provided evidence that the hermetic writings were of much later date than the Mosaïc period of origin previously ascribed to them.

Cassini, Gian Domenico (1625–1712): Italian astronomer and original member of the Royal Academy of Sciences in Paris.

Cavendish, Margaret (1623–1673): Duchess of Newcastle, English writer and anti-experimentalist, materialist philosopher.

Cesi, Federico (1585–1630): Italian founder of the Accademia dei Lyncei.

Cicero, Marcus Tullius (106–43 BC): Roman statesman and orator.

Clarke, Samuel (1675–1729): English supporter of Newton, Boyle lecturer, and disputant with Leibniz over Newton's natural philosophy.

Clavius, Christoph (1537–1612): German Jesuit mathematician at the Collegio Romano.

Commandino, Federico (1509–1575): Italian mathematician and translator of Archimedes.

Conway, Anne (1631–1679): English philosopher, friend of Henry More, student of cabalism and denier of Cartesian dualism.

Copernicus, Nicolaus (1473–1543): Polish anti-Ptolemaïc astronomer; wrote *De revolutionibus* (1543).

Cosimo II de' Medici (1590–1621): Grand Duke of Tuscany and Florentine patron of Galileo.

Cotes, Roger (1682–1716): English follower of Newton and editor of the *Principia*'s second edition (1713).

Croll, Oswald (*c*.1560–1609): German Paracelsian alchemist.

Dee, John (1527–1608): English mathematician and mystic.

Desaguliers, John Theophilus (1683–1744): Curator of Experiments for the Royal Society and popular Newtonian public science lecturer.

Descartes, René (1596–1650): French philosopher and mathematician.

Descartes, Catherine (1637–1706): niece of René and critic of mind–body dualism.

Digges, Thomas (1546–1595): English mathematician and early adherent of Copernicanism.

Diophantus of Alexandria (fl. *c*.250?): Greek mathematician whose work (the *Arithmetic*) stimulated the development of algebra.

Dioscorides (fl. 50–70): Greek botanist and physician.

Elizabeth of Bohemia (1618–1680): philosophical correspondent of Descartes, dedicatee of his *Passions of the Soul* (1647), and princess-daughter of the briefly reigning King of Bohemia Frederick V. Later an abbess.

Epicurus (341–270 BC): Greek philosopher and atomist.

Euclid (fl. *c*.295 BC): Greek mathematician, author of the *Elements*.

Ficino, Marsilio (1433–1499): Italian philosopher, Platonist, and translator of Plato and texts of Hermes Trismegistus.

Galen (129–*c*.200): Greek physician and anatomist.

Galileo Galilei (1564–1642): Italian astronomer, mathematician, and natural philosopher.

Gassendi, Pierre (1592–1655): French sceptical philosopher and reviver of Epicurean atomism.

Gilbert, William (1544–1603): English natural philosopher best known for his work on magnetism (*De magnete*, 1600).

Halley, Edmund (1656–1743): English astronomer and natural philosopher.

Harriot, Thomas (*c*.1560–1621): English mathematician.

Harvey, William (1578–1657): English physician and anatomist who argued for the circulation of the blood.

Hauksbee, Francis (*c*.1666–1713): English experimentalist, Curator of Experiments to the Royal Society in early years of Newton's presidency.

Hermes Trismegistus: mythical supposed Egyptian author of the so-called hermetic corpus of writings, which were thought to date from the time of Moses until Casaubon's work (q.v.).

Hobbes, Thomas (1588–1679): English philosopher and mathematician.

Hooke, Robert (1635–1702): English experimentalist; assistant to Robert Boyle (q.v.) in the 1650s, then the first Curator of Experiments to the Royal Society. Author of *Micrographia* (1665).

Huygens, Christiaan (1629–1695): Dutch mathematician and mechanical philosopher.

Huygens, Constantijn (1596–1687): Dutch diplomat and father of Christiaan.

Kepler, Johannes (1571–1630): German mathematician/astronomer.

Kircher, Athanasius (1602–1680): German Jesuit philosopher and polymath; spent most of his career in Rome.

Leibniz, Gottfried Wilhelm (1646–1716): German philosopher and mathematician.

Leopold de' Medici (1617–1675): Florentine noble founder of Accademia del Cimento; later a Cardinal.

Libavius, Andreas (*c*.1560–1616): German chemist.

Locke, John (1632–1704): English philosopher, author of *Essay Concerning Human Understanding* (1690).

Luther, Martin (1483–1546): German religious reformer and founder of Lutheranism.

Malebranche, Nicolas (1638–1715): French philosopher and follower of Descartes.

Mästlin, Michael (1550–1631): German astronomer and teacher of Kepler (q.v.); early Copernican.

Melanchthon, Philip (1497–1560): German follower of Martin Luther (q.v.) and Lutheran educational reformer.

Mersenne, Marin (1588–1648): French mathematician and chief correspondent of Descartes.

Mondino de' Liuzzi (*c*.1275–1326): Italian physician and anatomist; wrote standard digest of Galenic anatomy.

Monte, Guidobaldo dal (1545–1607): Italian nobleman, mathematician, and friend of Galileo.

More, Henry (1614–1687): English philosopher ("Cambridge Platonist").

Mydorge, Claude (1585–1647): French mathematician and friend of Descartes.

Newton, Isaac (1642–1727): English mathematician and natural philosopher; author of *Principia* (1687) and *Opticks* (1704).

Nicholas of Cusa (*c*.1401–1464): Cardinal in Catholic Church; philosopher and proposer of an infinite universe.

Oldenburg, Henry (*c*.1618–1677): Expatriate German in England, first secretary of the Royal Society and prodigious philosophical correspondent.

Osiander, Andreas (1498–1552): German Lutheran theologian who wrote anonymous preface to Copernicus's *De revolutionibus* (1543).

Paracelsus (*c*.1493–1541): Swiss medical reformer and mystic.

Pascal, Blaise (1623–1662): French mathematician.

Peucer, Caspar (1525–1602): German astronomer who used Copernicus's *De revolutionibus* at Lutheran university at Wittenberg.

Peurbach, Georg (1423–1462): German Ptolemaïc astronomer; wrote *Theoricae novae planetarum*.

Piccolomini, Alessandro (1508–1578): Italian philosopher who denied scientific status to mathematics.

Pico della Mirandola, Giovanni (1463–1494): Italian neo-Platonist.
Plato (*c.*427–347 BC): Greek teacher of Aristotle and the original rationalist philosopher; taught that mathematics is important in natural philosophy.
Pliny the Elder (*c.*23–79): Roman author of the *Natural History*.
Plutarch (*c.*46–*c.*120): Roman biographer and gossip.
Power, Henry (1623–1668): English natural philosopher and author of *Experimental Philosophy* (1664).
Ptolemy, Claudius (*c.*100–*c.*170): Greek astronomer.
Pyrrho of Elis (*c.*360–270 BC): Greek sceptical philosopher; founder of Pyrrhonism.
Rambouillet, marquise de (1588–1665): French hostess of first major Parisian salon.
Ray, John (1627–1705): English naturalist.
Recorde, Robert (*c.*1510–1558): English mathematical practitioner.
Redi, Francesco (1626–1697): Italian physician and zoologist; member of Accademia del Cimento.
Regiomontanus, Johannes (1436–1476): German humanist mathematician/astronomer.
Régis, Pierre Sylvain (1632–1707): French Cartesian lecturer.
Reinhold, Erasmus (1511–1553): German astronomer at University of Wittenberg; produced Prutenic tables based on Copernicus's *De revolutionibus*.
Reuchlin, Johannes (1455–1522): German Christian Cabalist.
Rhazes [Al-Razi] (865–925): Arabic medical writer.
Rheticus, Georgius (1514–1574): German mathematician. Disciple and publicist of Copernicus.
Ricci, Matteo (1552–1610): Italian Jesuit missionary to China.
Riccioli, Giambattista (1598–1671): Italian Jesuit astronomer at Bologna.
Rømer, Ole (1644–1710): Danish astronomer; original member of Royal Academy of Sciences.
Rouhault, Jacques (1620–1672): French Cartesian lecturer.
Scheiner, Christoph (1573–1650): German Jesuit astronomer.
Scudéry, Madeleine de (1607–1701): French salon hostess.
Sextus Empiricus (fl. *c.*200): Greek sceptic and follower of Pyrrho of Elis (q.v.).
Socrates (*c.*470–399 BC): Greek moral philosopher, teacher of Plato.
Sprat, Thomas (1635–1713): English Fellow of the Royal Society and author of *History of the Royal Society* (1667).
Tournefort, Joseph Pitton de (1656–1708): French naturalist.
Tycho Brahe (1546–1601): Danish astronomer, noted for his precise observational work.
Urban VIII (1568–1644): Pope, elected 1623 (previously Maffeo Barberini); gave Galileo the impression that the latter could once again speak openly about Copernicanism.
Valla, Lorenzo (*c.*1406–1457): Italian humanist.
Van Helmont, Johannes Baptista (1579–1644): physician and alchemical philosopher in the Spanish Netherlands (now Belgium).
Vesalius, Andreas (1514–1564): of Brussels; physician, surgeon, and anatomist, author of *De humani corporis fabrica* (1543).
Viète, François (1540–1603): French mathematician and developer of algebra.
Whiston, William (1667–1752): English Newtonian philosopher and mathematician.
Wilkins, John (1614–1672): English mathematician, one of the founders of the Royal Society.
Witelo (*c.*1230–after *c.*1275): Polish writer on optics.
Wren, Christopher (1632–1723): English mathematician and architect, member of Royal Society.
Wright, Edward (1561–1615): English mathematical practitioner.

Glossary of Major Terms

absolutism: a political ideal or arrangement in which all power in the state ultimately resides in the monarch; there are no independent sources of authority.

Accademia dei Lincei: a natural-philosophical society to which Galileo belonged; the name signifies "Academy of the Lynx-Eyed."

Accademia del Cimento: a private "academy" of experimenters founded in Florence by Prince Leopold of Tuscany in 1657; published its *Saggi di naturali esperienze* ("Essays of Natural Experiments") in 1667.

aether: originally, the Aristotelian matter composing the heavens; subsequently also applied by analogy to the matter of the Cartesian heavens, or to any very subtle, invisible form of matter.

alchemy: the esoteric study of matter and its qualitative changes, chiefly as brought about by the action of heat, and directed towards the purification of matter as represented by the creation of gold from lesser ("base") metals. The purification carried with it spiritual connotations, such that the state of the alchemist's soul was of relevance to the successful accomplishment of the goal.

Almagest: Ptolemy's great astronomical work that defined astronomy in the Islamic world and in Christian Europe until Copernicus.

Aristotelianism: a style of philosophy based on the writings of the ancient Greek philosopher Aristotle, and incorporating some of the central elements of his approach to knowledge.

atomism: the philosophical doctrine that the ultimate constituents of all matter are indivisible corpuscles whose properties serve to determine those of the bodies composed of them.

Averroïsm: a form of Aristotelianism due to the commentaries on Aristotle of the Arabic philosopher Averroës. It was a "fundamentalist" interpretation of Aristotle's philosophy that left no room for compromise with religious doctrine.

Cabalism: an occult philosophy, of Jewish origin, which held that the Hebrew words for things encoded profound knowledge discoverable through correct manipulation of their Hebrew letters.

Cartesianism: a strain of philosophy owing its central tenets to René Descartes. Descartes's main doctrines as discussed in the seventeenth century concerned his mechanistic explanations of physical phenomena as well as his arguments for the separability of the human mind from the body.

Cartesian dualism: Descartes's position that the mind and body are entirely distinct kinds of thing.

Collegio Romano: the flagship college of the early-modern Jesuit college system, located in Rome.

contextualism: a modern historiographical term used to designate attempts to understand the history of ideas by reference to the social and political contexts in which those ideas were promoted.

Copernicanism: Copernicus's doctrine that the earth orbits around a stationary sun once a year, or an adherence to the geometrical models given in Copernicus's *De revolutionibus* for calculating celestial appearances.

corpuscular; corpuscularism: referring to a view of matter as composed of minute particles, regardless of whether these are in principle divisible (a "corpuscle" is, literally, a "little body"); cf. "atomism."

cosmology: the philosophy of the universe as a whole and its structure; the physics of the heavens.

empiricism: a philosophical stance that holds that all knowledge is rooted in the senses and the experience that they provide.

Enlightenment: a term describing a dominant philosophical and cultural movement in eighteenth-century Europe that stressed the power of reason and experience in establishing reliable and sound knowledge, venerated the seventeenth-century Englishmen Isaac Newton and John Locke, and that saw such reason as a weapon against superstition and the political power of entrenched traditional authorities, including the Church.

experimental philosophy: a term used by Robert Boyle and other Fellows of the early Royal Society to refer to a natural philosophy that relied on gathering facts from experimental and observational work.

geocentric: centred on the earth; used in astronomy.

heliocentric: centred on the sun; used in astronomy.

hermeticism: the doctrines promulgated in the writings of the hermetic corpus, supposed to date from distant antiquity; matter was held to be alive and occult sympathies ran through the universe. See also "Hermes Trismegistus."

Holy Roman Empire: a loose political union of central European, mostly German, states. Its head was the Holy Roman Emperor, who was elected by the rulers of the more important of the constituent states.

humanism: a cultural movement originating in Italy in the late fourteenth century and the fifteenth century. It consisted of a reverence for and close study of the writings of Greek and Roman antiquity, and promoted attempts at the emulation of ancient cultural achievements. Educationally, it involved a stress on the teaching of classical rhetoric.

induction: a term from classical logic and rhetoric, used by Francis Bacon to mean a process of inference based on an exhaustive collection of empirical facts, and by Isaac Newton to refer to the generalization of properties from one experiment to all situations deemed similar to it.

Jesuits: the intellectual élite of the Roman Catholic Church in the later sixteenth century and the seventeenth century. The Jesuits ran a network of prestigious colleges throughout Catholic Europe; many Jesuits were prominent practitioners of the various mathematical sciences, as well as of branches of natural philosophy.

materia medica: something from which medical remedies can be prepared.

mathematics: in this period, a general term referring both to "pure" mathematics and "mixed" mathematics. The first category included geometry, arithmetic, algebra and (by the end of the seventeenth century) the calculus, the last two known collectively as "analysis." The second category included all studies that involved the use of quantity and the techniques of the "pure" branches in studying actual, non-abstract things in the world, especially mathematical astronomy, music theory, mechanics, and geometrical optics. Astronomers were often called "mathematicians."

mechanical philosophy: a term coined by Robert Boyle to describe any general explanatory system of the physical world that treated its phenomena as due to nothing but pieces of inert matter interacting with one another by virtue of their shapes, sizes, and motions.

mechanism: a term stressing the explanatory ideal of the "mechanical philosophy."

metaphysics: that branch of knowledge which considers the fundamental categories of reality, such as existence, being, matter, space, etc.

micro-mechanisms: explanations for phenomena which posit submicroscopic mechanical arrangements of material parts.

natural history: a descriptive account of things in the physical (non-human) world; particularly, but by no means exclusively, applied to systematic description of plants and animals.

natural philosophy: a category, also know as "physics," approximately equal to Aristotle's term *physis*. It referred to systematic knowledge of all aspects of the physical world, including living things, and in the sixteenth and seventeenth centuries routinely understood that world as being God's Creation. It therefore possessed strong theological implications.

neo-Platonism: a philosophy deriving from late-antique followers of Plato such as Plotinus or Proclus. It stresses Plato's praise of mathematics as a means of knowing the world, and transforms it into a kind of mathematical mysticism.

Newtonianism: a style of philosophy, first developing in England in the 1690s, that claimed to follow the doctrines of Isaac Newton regarding the right way to learn about nature (empiricism and induction; mathematics) and the content and structure of the physical universe.

occult: literally, "hidden." A term used in Aristotelian philosophy to refer to inaccessible and presumably unknowable causes of evident phenomena, such as magnetism.

operationalism: a philosophical ideal whereby the truth of a statement is shown by the possibility of putting it to practical use (to *work*).

Paracelsianism: the medical philosophy of Paracelsus and his followers, which stresses occult sympathies between various parts of the world as the key to curing ailments.

philosopher: in the early-modern period, a term with much wider scope than nowadays. A philosopher could be anyone who thought about and sought knowledge in any area; as, natural philosopher, moral philosopher, political philosopher. Rather like the modern term "intellectual."

physico-mathematics: a coinage of the seventeenth century, indicating the use of mathematics in the study of physical things, and usually carrying the implication that mathematical understanding could provide knowledge of the physical *causes* of phenomena. Cf. "mathematics."

physics: a general term for the study of the natural world, whether animate or inanimate. A practical synonym for "natural philosophy."

Pyrrhonism: A form of philosophical scepticism ascribed to Pyrrho of Elis, and promulgated by Sextus Empiricus. It held that nothing can be known with certainty, and that we should therefore suspend judgement regarding all truth-claims whatsoever.

rationalism: a philosophical stance that holds that the key to knowledge is the correct use of reason; that we learn truths by reasoning our way to them.

Renaissance: the historical period from 1400 or so to around 1600, depending on the particular region of Europe. The word means "rebirth," and refers to the period in which high culture devoted itself to the recovery of the civilization of classical antiquity. Cf. "humanism."

Royal Academy of Sciences (Académie Royale des Sciences): founded in 1666 as an arm of the French state. Its restricted and paid membership conducted inquiry into mathematical studies such as astronomy and navigation, and natural-philosophical ("physical") inquiries such as chemistry and zoology.

Royal Society: the Royal Society of London for the Improving of Natural Knowledge was established in the early 1660s. Its Fellows committed themselves to self-described Baconian experimental and natural historical inquiry, with a stress on the practical usefulness of natural philosophy.

scepticism: the general philosophical position that denies or calls into absolute doubt all claims to truth.

scholasticism; scholastic: scholasticism is a term applied to the intellectual and academic style of the medieval universities, a style stressing debate, disputation, and the effective use of canonical texts (such as those of Aristotle) in the making of arguments. A "scholastic" is a practitioner of this style.

scholastic Aristotelianism: Aristotelian philosophy pursued according to scholastic procedures.

scientia: the Latin translation of the Greek "epistēmē." Demonstrable, certain knowledge, as contrasted with *opinio*, "opinion."

Stoic; Stoicism: An ancient Greek philosophical school founded by Zeno of Citium, which propounded ethical and natural-philosophical doctrines of considerable influence in the sixteenth and seventeenth centuries. Stoic physics regarded matter as active and self-moving, and space as being filled with a fluid substance (*pneuma*) which served to connect all parts of the universe to all others.

syllogism: the central technical device in formal logic in the universities of the Middle Ages and early-modern period, derived from Aristotle's writings on logic, and consisting of a "major premise" (all As are B), a "minor premise" (C is A), and a "conclusion" (therefore C is B).

Thomism: a philosophical approach based on the work of St Thomas Aquinas.

vita activa: the "active life," a mode of living that involves engagement in the world, recommended by many humanists, including Lorenzo Valla.

vita contemplativa: the "contemplative life," recommended as best by Aristotle, in which one withdraws from society to pursue self-improvement through exercise of the intellect.

Index